49 Advances in Biochemical Engineering Biotechnology

Managing Editor: A. Fiechter

Chromatography

Volume Editor: George T. Tsao

With contributions by
P. M. Boyer, T. Gu, J. T.-A. Hsu,
C. M. Ladisch, M. R. Ladisch, M.-J. Syu,
G.-J. Tsai, G. T. Tsao, A. Velayudhan,
Y. Yang

With 59 Figures and 15 Tables

Springer-Verlag
Berlin Heidelberg GmbH

ISBN 978-3-662-14955-3   ISBN 978-3-540-47513-2 (eBook)
DOI 10.1007/978-3-540-47513-2

Library of Congress Catalog Card Number 72-152360

Originally published by Springer-Verlag Berlin Heidelberg New York in 1993
Softcover reprint of the hardcover 1st edition 1993

Typesetting: Macmillan India Ltd., Bangalore-25

02/3020 - 5 4 3 2 1 0 - Printed on acid-free paper

Attention all "Enzyme Handbook" Users:

A file with the complete volume indexes Vols. 1 through 5 in delimited ASCII format is available for downloading at no charge from the Springer EARN mailbox. Delimited ASCII format can be imported into most databanks.

The file has been compressed using the popular shareware program "PKZIP" (Trademark of PKware INc., PKZIP is available from most BBS and shareware distributors).

This file distributed without any expressed or implied warranty.

To receive this file send an e-mail message to:
SVSERV@DHDSPRI6.BITNET.
The message must be: "GET/ENZHB/ENZ_HB.ZIP".

SPSERV is an automatic data distribution system. It responds to your message. The following commands are available:

HELP	returns a detailed instruction set for the use of SVSERV,
DIR (*name*)	returns a list of files available in the directory "name",
INDEX (*name*)	same as "DIR"
CD *<name>*	changes to directory "name",
SEND *<filename>*	invokes a message with the file "filename"
GET *<filename>*	same as "SEND".

Table of Contents

Protein Purification by Dye-Ligand Chromatography

Philip M. Boyer and James T. Hsu*
Bioprocessing Institute, Department of Chemical Engineering,
Lehigh University, Bethlehem, Pennsylvania 18015, USA

Dye-ligand chromatography has developed into an important method for large-scale purification of proteins. The utility of the reactive dyes as affinity ligands results from their unique chemistry, which confers both the ability to interact with a large number of proteins as well as easy immobilization on typical adsorbent matrices. Reactive dyes can bind proteins either by specific interactions at the protein's active site or by a range of non-specific interactions. Divalent metals participate in yet another type of protein-reactive dye interactions which involve the formation of a ternary complex. All of these types of interactions have been exploited in schemes for protein purification. Many factors contribute to the successful operation of a dye-ligand chromatography process. These include adsorbent properties, such as matrix type and ligand concentration, the buffer conditions employed in the adsorption and elution stages, and contacting parameters like flowrate and column geometry. Dye-ligand chromatography has been demonstrated to be suitable for large-scale protein purification due to their high selectivity, stability, and economy. Also, the issue of dye leakage and process validation of large-scale dye-ligand chromatography has been discussed. Reactive dyes have also been applied in high performance liquid affinity chromatographic techniques for protein purification, as well as non-chromatographic techniques including affinity partition, affinity membrane separations, affinity cross-flow filtration, and affinity precipitation.

* Author to whom correspondence should be addressed

Advances in Biochemical Engineering
Biotechnology, Vol. 49
Managing Editor: A. Fiechter
© Springer-Verlag Berlin Heidelberg 1993

List of Symbols

C Solute concentration in bulk liquid, mg cm^{-3}

C_p Solute concentration in pore liquid, mg cm^{-3} pore

C_s Adsorbed solute concentration, mg cm^{-3} solid

C^* Solute concentration in bulk liquid at equilibrium, mg cm^{-3}

C_s^* Adsorbed solute concentration at equilibrium, mg cm^{-3} solid

D_L Axial dispersion coefficient, cm^2 s^{-1}

D_p Pore diffusivity of the solute, cm^2 s^{-1}

K_D Dissociation constant in Langmiur isotherm model, mg cm^{-3}

k Capacity parameter in Freundlich isotherm model, mg cm^{-3} solid

k_F Film mass transfer coefficient, cm s^{-1}

k_1 Intrinsic forward interaction rate constant, cm^3 mg^{-1} s^{-1}

k_2 Intrinsic reverse interaction rate constant, s^{-1}

n Exponential parameter in Freundlich isotherm model

Q_m Maximum binding capacity in Langmuir isotherm model, mg cm^{-3} solid

R Adsorbent particle radius, cm

r Radial distance from center of particle, cm

t Time, s

V Average interstitial velocity of mobile phase in column, cm s^{-1}

z Axial distance along column from inlet, cm

ε Column void fraction

ε_p Intraparticle void fraction

1 Introduction

The successful exploitation of biotechnology depends on the development of efficient and reliable methods for protein purification. Biotechnological innovations have made their mark on a number of diverse fields including the pharmaceutical, biomedical, agrochemical, and food processing industries. While these innovations are in large measure a result of recent advances in recombinant DNA and cell culture technologies, none would have been possible without concomitant advances in protein purification technologies. Nowhere is the need for protein purification methods as great as in the pharmaceutical industry where many of the new products are proteins: vaccines, blood proteins, and hormones. The purity requirements for these products are very exacting and it is not uncommon for purification and validation to represent the majority of the development cost of a new therapeutical protein. Another area of increasing importance is that of monoclonal antibodies production. Developments in this area have led to new technologies such as clinical diagnostics, cancer therapy, tumor imaging, and immunoaffinity chromatography. In each of these applications, protein purification, often to homogeneity, is a key step in the commercial realization of biotechnology.

Chromatography has been established as the pre-eminent method for the purification of proteins and other biomolecules. While there are many reasons for its widespread use, the rich variety of adsorbents and the ability to resolve closely related proteins are two important factors which contribute to its popularity. Gel filtration, ion exchange, and hydroxyapatite chromatography were the first types of chromatography routinely used for protein purification in the late 1950s and 1960s [1]. These methods made possible the purification of a large number of proteins, consequently research on protein purification greatly expanded in this period. After the introduction in 1967 of the CNBr activation technique [2] biospecific affinity chromatography was rapidly developed. These affinity matrices attempted to take advantage of a protein's extremely specific biological function as a means for its isolation. Early in the development of biospecific affinity chromatography, it was predicted that all other methods would become outdated [3, 4]. However, a number of unforeseen problems have been encountered in the application of affinity chromatography which have prevented its widespread use at the process-scale [5-7]. These problems include the unavailability and expense of the biological ligands, the lack of biological stability and activity of the immobilized ligand, ligand leakage, and low binding capacity.

In response to these problems, some researchers investigated the development of group-specific adsorbents based on immobilized coenzymes or nucleotides [8, 9]. Although this technique resulted in decreased development cost and time, the problems of low capacity and stability, along with the high cost of these cofactors still limited large scale application. As a consequence, interest in adsorbents which employ synthetic ligands which exhibit empirical, rather than

biospecific, affinity for proteins have grown [6]. These ligands may be characterized as inexpensive molecules which interact with proteins by a range of forces.
Dye-ligand chromatography is perhaps the best example of the application of
such empirical adsorbents. The ligands employed in dye-ligand chromatography are reactive dyes which were developed for use in the textile industry.

Dye-ligand chromatography has received considerable attention as method
of protein purification because of its many advantages over other forms of
affinity chromatography. Some of these advantages are listed in Table 1. The
low cost and general availability of the reactive dyes, along with the ease of
immobilization, represent major advantages over other biospecific affinity
ligands such as antibodies, enzymes, cofactors or substrates.

The immobilization of reactive dyes is much simpler than methods involving
CNBr-activated matrices due to the relatively mild conditions which are
involved. In addition, the covalent linkage formed in dye attachment is much
more stable than the isouronium linkage typically used for ligand immobilization [10–12]. The stability of immobilized reactive dyes allows them to be
stored for several years with no loss of activity [13]. Furthermore, these
adsorbents can withstand the harsh conditions employed in regeneration
procedures; therefore, they are suitable for repeated use, as demonstrated by
More et al. [14] who reported the use of a Procion Blue HB-Sepharose column
for more than 100 cycles without noticeable loss of capacity or specificity. These
factors, coupled with the ever-increasing number of proteins which may be
purified, have established dye-ligand chromatography as an important method
for protein purification at both laboratory and production scale.

The rise of the use of reactive dyes from their serendipitous discovery to their
acceptance as affinity ligands today has been an interesting one. The discovery
that reactive dyes interact with proteins resulted from the fact that Blue
Dextran, which consisted of the reactive dye Cibacron Blue F-3GA coupled to
dextran, was widely used as a void marker in gel filtration chromatography. It
was found that phosphofructokinase and pyruvate kinase eluted in the void
volume when co-chromatographed with Blue Dextran, while they appeared
much later when the void marker was excluded [15, 16]. This discovery was

Table 1. Advantages of reactive dyes as affinity ligands for protein purification

Economical and widely available

Ease of immobilization

Avoids hazardous and toxic reagents in matrix activation

Stable against biological and chemical attack

Storage of adsorbent without loss of activity

Reusable: withstands cleaning and sterilization

Ease of scale-up

High capacity

Medium specificity

soon applied for the purification of these enzymes. It was soon determined that Cibacron Blue F-3GA, and not the Dextran, was responsible for the interaction, and procedures using directly coupled to polysaccharide matrices were developed [17]. Throughout the first ten years, Cibacron Blue F-3GA was the only dye seriously considered, and studies focused on identifying why this molecule interacted so specifically with dehydrogenase and kinase enzymes. Reports showed that substrates competitively inhibited of dye interactions, which led to claims that the molecule bound at the active site [17–19]. These claims were soon confirmed by X-ray crystallographic studies, which showed Cibacron Blue F-3GA interacted with the nucleotide-binding domain of horse liver alcohol dehydrogenase [20]. It was further proposed that immobilized Cibacron Blue F-3GA could be used as a probe for the presence of the dinucleotide fold in a protein [21, 22]. This theory soon began to unravel as more and more examples, such as arabinose binding protein [23] and albumin [24] were reported which contradicted it. Protein-binding by Cibacron Blue F-3GA was beginning to be viewed as a combination of several forces including both electrostatic and hydrophobic interactions [25].

This new understanding opened the door to investigations regarding the use of other reactive dyes in protein purification schemes. Procion Red HE-3B was one of the first dyes to be recognized as having similar protein-binding capability [13, 19]. However, several studies showed that many reactive dyes could interact with proteins and consequently may serve as ligands for their purification [26–30]. Screening procedures based either on the free dye interactions [27–29] or on small beds of the dye adsorbents [31–33] were used to identify suitable ligands. Along with the notion that many reactive dyes may be useful in protein purification came new ideas about contacting schemes for using these adsorbents. Perhaps the most interesting was tandem-dye chromatography [31], which employed a "negative" column to remove contaminating protein from the extract before it was fed to a typical "positive" column. The biggest challenges in the development of this type of system is the identification of suitable ligands [33], and suitable operating conditions for the adsorption and elution stages of the process [34]. Through these developments, dye-ligand chromatography was transformed from a curiosity involving one or two dyes, to a distinct class of protein chromatography that has unique challenges in adsorbent screening and process optimization.

Recent work in the use of immobilized reactive dyes has been in directions almost as numerous as there are reactive dyes themselves. Fundamental research into dye-protein interactions continues to be investigated. One interesting development in this area is the rational modification of reactive dyes through three-dimensional computer simulation in order to increase the biomimetic character of the dye-protein interaction. The manufacture of new materials has also renewed interest in the support matrix used in dye-ligand chromatography, and there have been numerous reports of the use of HPLC dye-ligand adsorbents. Reactive dyes are also finding use in non-chromatographic applications such as affinity membrane separations, affinity cross-flow ultrafiltration, affinity biphasic partitioning and affinity precipitation. Dye-ligand chromatography is also beginning to be applied in large-scale protein purification, where process validation and dye-leakage become important issues.

2 Reactive Dye Chemistry

An important objective in dyestuff manufacture is to produce dyes which will not be laundered out of the material. By the 1950s, colorfast dyes had been designed for the synthetic and animal-derived fibers used in the textile industry. Cotton, however, presented significant problems due to its cellulosic structure, which rendered the dyes designed for these other materials ineffective. The reactive dyes were introduced in 1956 as colorfast dyes for use with cellulosic fibers [35]. These dyes were large hydrophilic molecules which contained a reactive group in addition to the chromophore. The reactive group was designed to react covalently with the cellulosic fibers under mild conditions, this covalent linkage is responsible for the high fastness of these dyes. Due to the wide variety of the applications requiring these dyes in the textile industry a plethora of dyes have been produced [36]. This discussion of reactive dye chemistry is intended simply to acquaint the reader with the principle structural components common to all reactive dyes, and the methods used to immobilize these compounds for use in dye-ligand chromatography.

2.1 Reactive Dye Structure and Chemistry

As mentioned above, reactive dyes consist of two major components: a chromophore attached to a reactive group. By far the most commonly used reactive groups are the monochloro- and dichloro-triazine rings. The structure of several typical triazinyl dyes are shown in Fig. 1. These dyes are prepared by the reaction of cyanuric chloric with an amino-containing dye, thereby producing a dichlorotriazinyl dye, which corresponds to the Procion MX range of dyes produced by ICI. As seen in Fig. 1 the triazine ring in these dyes contain two labile chlorine atoms which makes dyes of this type highly reactive. Subsequent reaction of these dyes with an amine or alcohol causes the replacement of one of the triazinyl chlorine atoms and produces a less reactive monochlorotriazine dye which corresponds to ICI's Procion H range and Ciba-Geigy's Cibacron range. Dyestuff manufacturers have developed a number of other reactive groups in order to generate a wide range of reactivities for these textile dyes [5]; however, with the exception of the Remazol range of dyes made by Hoechst, few of these dyes have found application in dye-ligand chromatography.

Reaction of chlorotriazine dyes with cellulose or other carbohydrates is accomplished by the following process [36]. The carbohydrate substrate (cellulose or other matrix) is suspended in an aqueous solution of the dye. Salt is

Procion Blue MX·3G

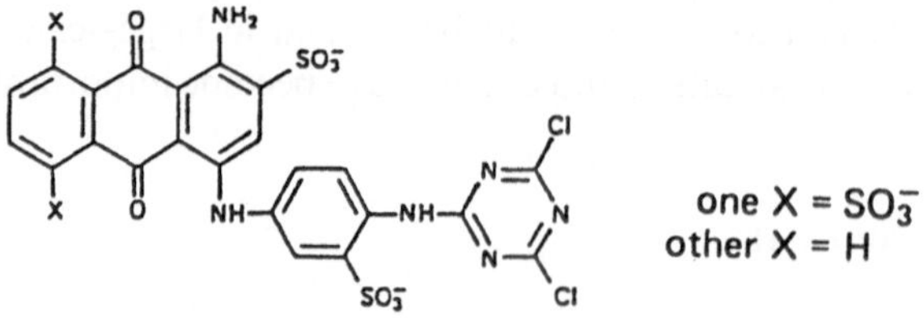

one X = SO$_3^-$
other X = H

Procion Blue H-B

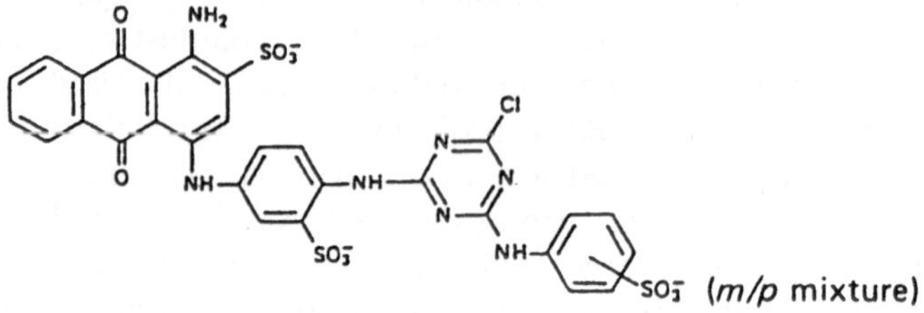

(*m/p* mixture)

Procion Red H-3B

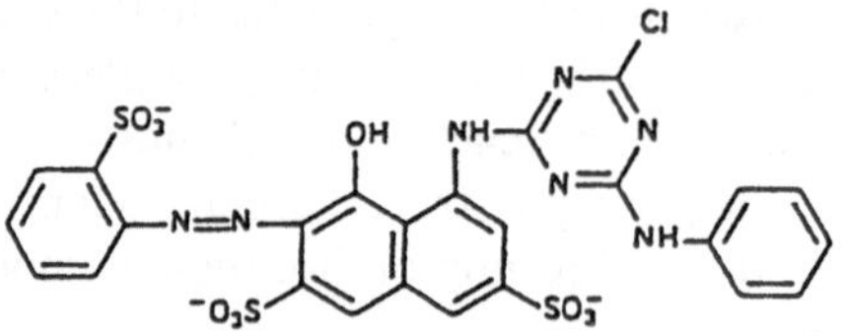

Procion Red HE-3B

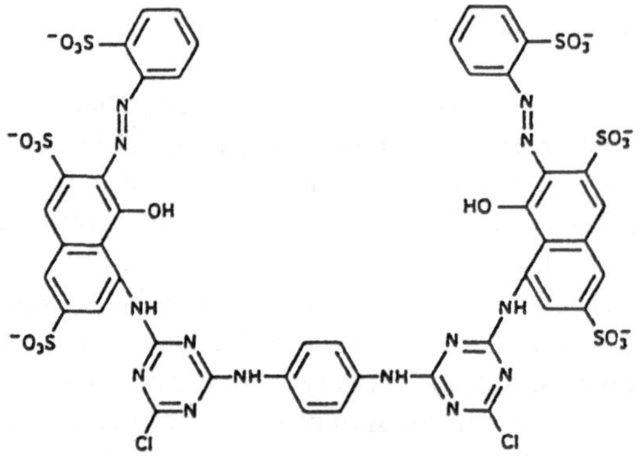

Procion Yellow H-A

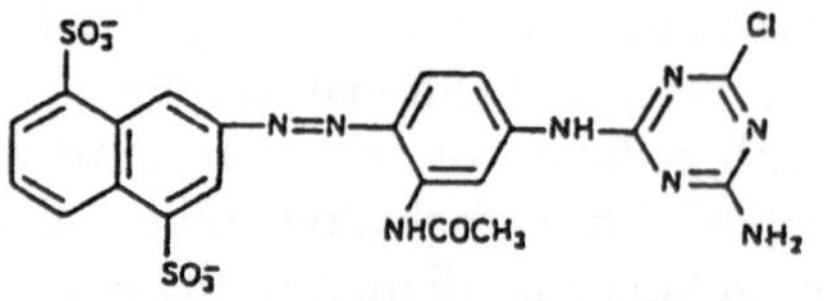

Fig 1. Chemical structure of several typical reactive dyes [186]

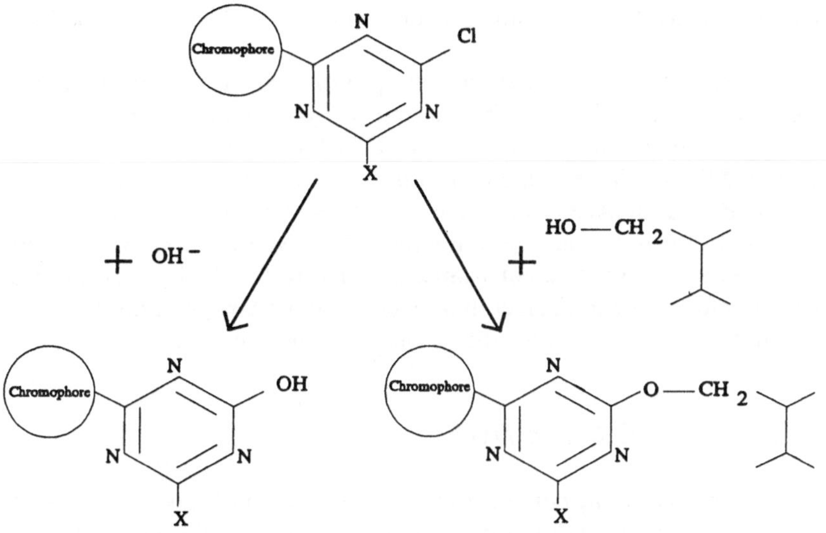

Fig 2. Nucleophilic attach of a triazene dye by either carbohydrate-OH⁻ or hydroxyl ion

added to cause the dye to adsorb onto the surface of the substrate. When adsorption is complete, the pH of the solution is raised by adding either carbonate or hydroxide. As seen in Fig. 2, at this elevated pH the reactive chlorine atom on the triazine ring is subject to nucleophilic attack by either free hydroxyl ions or substrate hydroxyls. If attacked by a substrate-hydroxyl, then the dye is immobilized on the substrate; attack by a free hydroxyl ion, however, results in hydrolysis and the dye is lost to the dyeing process. Reaction of the dye with the substrate is favored because of two factors: the close proximity of the dye and carbohydrate fiber due to adsorption of the dye and the high ratio of substrate hydroxyls relative to free hydroxyl ions within the substrate [37]. The high reactivity of the dichlorotriazine dyes allow this dyeing process to be completed within 1 h under mild conditions, typically at a temperature between 30 and 40 °C and at a pH 10.5. The monochlorotriazine are much less reactive and therefore require more extreme conditions, usually at a temperature of 85 °C and a pH of 11. The Remazol dyes use a reactive system which is based on an activated double bond. As before, a substrate hydroxyl attacks the dye, however in this case, the dye is bound as a vinyl sulphone [34, 36].

The chromophore in the reactive dye must exhibit certain properties in addition to its characteristic color [36]. The first requirement is that the chromophore contain enough sulfonate groups so that it is soluble in water, since the dyeing process is to be conducted in that medium. Secondly, it must contain an amino group to which the reactive group may be attached. While these requirements do not impose any great restrictions on the choice of the chromophore, in practice, the dyes developed at ICI are derived from three chromophores employed and belong to one of three groups: azo, anthraquinone, and phthalocyanine [5]. These chromophores, used singly, in combination, or in metal complexes provide a spectrum of shades which cover the range from black to yellow.

There are several possible naming conventions which can be used for reactive dyes [34, 36]. A dye may be referred to by its color index name (i.e. Reactive Blue 4), color index number (i.e. CI 61205), or by a trade name (Procion Blue MX-R). The naming convention has become very confusing for the Cibacron Blue F-3GA/Procion Blue H-B dyes. The two dyes have been given the same Color Index name and number (Reactive Blue 2, CI 61211), but have been shown to have different isomeric structure [5]. This situation has been further confused due to the fact that several commercial preparations have names which are inconsistent with their isomeric structure [38].

2.2 Purification of Reactive Dyes

All commercial preparations contain various additives which give the product consistency and promotes stability of the reactive dye [36, 39]. The dichloro-triazine dyes are especially susceptible to hydrolysis, and therefore need to be stored in the presence of buffer salts in order to minimize this loss. Sodium chloride is often added to the dyes in order to dilute the dye content to a standard value and trace quantities of surfactants are often added to prevent dustiness in the dye preparation. Commercial dyes will also contain small amounts of synthetic intermediates from the final reaction stages, as well as some hydrolyzed dye. While these contaminants have no adverse effect on the dyeing process, they may affect analytical biochemical studies. In this regard, it has been reported that inhibition of glycerate kinase and isoleucyl-tRNA synthetase results from impurities in the commercial Cibacron Blue F-3GA, rather than by the dye itself [40].

The effects of contaminants on protein purification by dye-ligand chromatography are not expected to be great, because few of the contaminants will be immobilized on the support matrix, and proper washing of the matrix should remove adsorbed contaminants. Thus, as suggested previously [39], purification of reactive dyes is necessary only when conducting analytical studies with the free dye. In cases where immobilized reactive dyes are used, purification of the dye before immobilization is not likely to be necessary. Reactive dyes have been purified by a number of chromatographic procedures such as thin-layer chromatography, high performance liquid chromatography and column chromatography on silica gel or Sephadex LH-20 [5].

2.3 Immobilization of Reactive Dyes on Chromatographic Matrices

The most commonly used matrices for dye-ligand chromatography are gel filtration media including cross-linked agarose, cross-linked dextran, and beaded cellulose. Cross-linked agarose appears to be the best "general purpose" matrix, due to its structural stability, flow properties, low incidence of non-specific adsorption, and open pore structure which allows high protein binding capacity [39]. The dye-matrices are typically prepared with the dye immobilized directly, rather than indirectly via a spacer arm, because of significant advantages in ligand leakage, capacity, and simplicity of immobilization.

A standard method for the immobilization of reactive dyes directly via the triazine ring onto carbohydrate matrices has been described [39]. This method has also been adapted for the large-scale preparation of dye-ligand adsorbents [30] on cross-linked agarose. In the laboratory scale method [39], 5 g of the polysaccharide matrix is washed with distilled water extensively on a sintered glass filter with slight suction. Extraneous water is suctioned from the matrix, and the moist cake is transferred to 5 ml of dye solution containing 50 mg reactive dye. The suspension is agitated for 30 min to allow the dye to diffuse completely into the matrix particles. One ml of a 22% solution of sodium chloride is added to raise the salt concentration to 2%, which causes the dye molecules to adsorb onto the matrix. After about 45 min when the adsorption is nearly complete, the reaction is initiated by addition of sodium hydroxide. Due to differences in reactivities, the sodium hydroxide concentration is adjusted to 0.01 N for dichlorotriazine dyes and 0.01 N for monochlorotriazine dyes. The reaction of dichlorotriazine dyes will generally be complete within 2–4 h at room temperature, whereas that of the less reactive monochlorotriazine will proceed over 3–5 h at room temperature. The reaction of monochlorotriazine dyes can be accelerated by raising the reaction temperature: at 37 °C adequate coupling required only two days, and at 60 °C only 2 h are required. Since few of the carbohydrate based matrices are stable at this high temperature, catalysts, such as tertiary amines, tetrazines, hydrazines or hydrazones, have added to accelerate the reaction of these monochlorotriazine dyes [5, 39, 41].

Other procedures advocate the use of 1% sodium carbonate rather than sodium hydroxide; [17, 42, 43], however, in the authors' experience, these conditions yield lower and less consistent dye immobilization than the above procedure. Optimal sodium hydroxide concentrations have been reported for dye immobilization on agarose: 0.01 M for dichlorotriazine dyes, and between 0.05 and 0.20 M for monochlorotriazine dyes [30]. The amount of dye immobilized on the matrix can be adjusted by varying the initial concentration of the dye solution; [27, 30]. It has been found that for Cibacron blue F-3GA immobilization on cross-linked agarose (6%), that a nearly linear relationship exists between initial dye concentration and the resulting immobilized dye concentration for initial concentrations in the range 0–15 μmole ml$^-$ [44], as shown in Fig. 3.

After the immobilization is complete, the unreacted dye must be washed from the adsorbent. Atkinson et al. [30] have recommended washing with copious amounts of the following series of solvents: (1) water, (2) 1 M sodium chloride in 25% ethanol, (3) water, (4) 1 M sodium chloride in 0.2 M phosphate, (5) water. Other solvents proposed for washing away unreacted dyes include water [45] 6 M urea in 0.5 M sodium hydroxide [34], 1–2 M sodium chloride

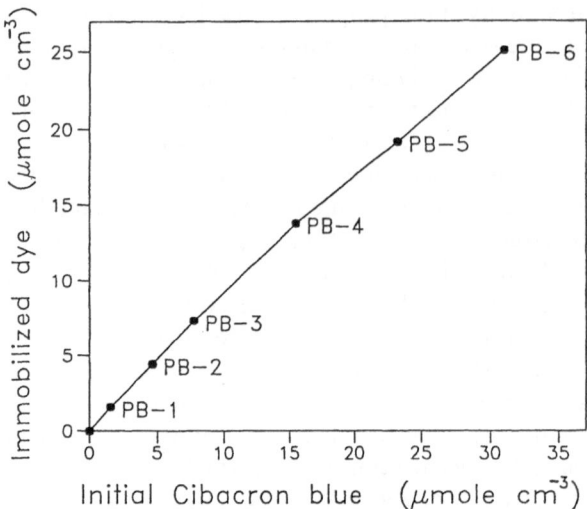

Fig 3. Immobilization of Cibacron Blue 3GA: dependence of immobilized ligand concentration on initial concentration of dye solution [44]

[39], and dimethyl sulfoxide (DMSO) [46]. This washing stage ensures that all unreacted dye is removed from the adsorbent before its use in protein purification, where dye leakage is often a major concern. After washing is complete, the dye should be stored in a neutral phosphate buffer at 4 °C. The addition of 0.02% sodium azide will prevent microbial contamination of the adsorbent [39].

The immobilized dye concentration can have important effects on protein adsorption, and therefore should be measured for each adsorbent prepared. The most accurate measurement of immobilized dye concentration is accomplished by dissolving the matrix and measuring the absorbance at the λ_{max} of the dye. Most carbohydrate gels may be dissolved in 6 M HCl at 37 °C, however, cross-linked agaroses and celluloses may require more extreme conditions for complete dissolution. After neutralization of the acid by the addition of sodium hydroxide, the absorbance may be measured. The dye concentration may then be calculated from the molar extinction coefficient, if it is known. Extinction coefficients for a number of dyes have been reported elsewhere [28, 47]. For those dyes for which the molecular structure has not been disclosed, the dye concentration may be reported in terms of mg ml^{-1} gel, however due to batch to batch variation in dyes, these values are less reliable [48].

3 Interactions of Reactive Dyes with Proteins

The various schemes which have been implemented to effect protein adsorption and elution in dye ligand adsorbents reflect the complex nature of the interactions between reactive dyes and proteins. Unlike the cases of ion exchange or hydrophobic interaction, where the basis for the protein-ligand interactions is understood reasonably well, protein adsorption behavior on immobilized reactive dyes cannot be predicted reliably; rather, it must be empirically measured. The study of these interactions has been, and very likely will continue to be, a primary focus of dye-ligand chromatography. It is well known that certain reactive dyes are structurally similar to nucleotide cofactors and have been shown to interact specifically with many enzymes that associate with these cofactors. However, many other proteins not associated with these cofactors also exhibit strong interactions with reactive dyes. Finally, the presence of certain metal ions have been observed to significantly alter protein binding by reactive dyes. Thus it is clear that no single mechanism can be responsible for all of these observations, so the various types of interactions will be discussed individually.

3.1 Specific Interactions

Since the reactive dyes are not naturally derived compounds, truly "biospecific" interactions with proteins are not possible, except for the possibility of interactions with antibodies raised against a dye. Nevertheless, these dyes have been found to exhibit highly selective interactions with many proteins. From the beginnings of dye-ligand chromatography, Cibacron Blue F-3GA was recognized for its high selectivity for nucleotide-dependent enzymes. It has been demonstrated through both X-ray crystallography and inhibition studies that this dye binds at the nucleotide-fold in many nucleotide-dependent enzymes [20, 28, 29]. Cibacron Blue F-3GA mimics the structure of nicotinamide adenine dinucleotide (NAD), especially in the planar ring structure and placement of negatively charged groups [20, 49]. Thus the selectivity of Cibacron Blue F-3GA for oxidoreductases, transferases and other enzymes which associate with nucleotide cofactors may be often a result of this specific interaction. Other reactive dyes have also been found to interact with enzymes at their active site [50, 51]. These cases each exhibit a "specific" interaction, in that the dye interacts at the protein's active site and competitive inhibition may be observed. However, it should also be noted that such specific interactions are likely to involve several types of interactive forces [52].

Understanding the fundamental interactions of reactive dyes with the active sites in proteins is the subject of many current investigations. One recent study has investigated the inhibition of luciferase, an ATP-dependent enzyme, by Cibacron Blue F-3GA, its fragments, and a steric isomer of the dye [53]. The chromophore appeared to be responsible for the specific interaction of the dye with this enzyme. Other researchers reached similar conclusions regarding the Cibacron Blue F-3GA interaction with horse liver alcohol dehydrogenase [38, 49]. Through computer simulation, based on the X-ray diffraction data for the Cibacron Blue F-3GA interaction with horse liver alcohol dehydrogenase, it was proposed that modification of the terminal ring of this dye would promote adsorption. Experimental studies of these modifications did demonstrate significant effects in binding by both free and immobilized dye analogues [54, 55]. Consideration of fundamental dye-protein interactions has also served as the basis for the design of cationic reactive dyes for the purification of proteolytic enzymes and proteins which interact with cationic substrates [46]. These studies represent a new approach to dye-ligand adsorbent design, founded on understanding the fundamental dye-protein interactions, with the aim of promoting the dye-protein interactions through alteration of the dye structure. The area of biomimetic ligand design may prove to be an important development in dye-ligand chromatography in the coming years.

3.2 Non-Specific Interactions

In contrast to the previous section, the adsorption of many enzymes and proteins results from interactions with immobilized reactive dyes at positions away from the active site. These cases are usually referred to as "non-specific" adsorption, even though the affinity of the interaction may be very high. The complex structure of the reactive dyes allows a wide range of interactions with proteins. The sulfonate moieties, introduced in the chromophore in order to increase solubility, can behave as cation exchangers. The chromophore of the dye is usually aromatic, and sometimes heterocyclic, which gives the reactive dyes some degree of hydrophobic character. Furthermore, reactive dyes can have hydrogen bonds and Van der Waals interactions with proteins. Thus it is clear that reactive dyes are multifunctional ligands, and that any or all of these forces may be involved in the adsorption of a protein [6].

These interactions are affected by the pH and ionic strength of the buffer. In general, protein binding from crude mixtures is observed to decrease when either the pH or ionic strength of the buffer is raised. These trends point to the importance of cation exchange in protein adsorption, however as seen in Fig. 4, cation exchange cannot account entirely for the observed behavior. In these systems, the protein adsorption very likely involves many types of interactions or even mixed interactions.

Recent studies have reported the adsorption of several model proteins in order to determine the types of forces which are important in these dye-protein interactions. In the first study, equilibrium adsorption of bovine serum albumin and lysozyme on Cibacron Blue F-3GA-Sepharose were determined for various conditions of pH and ionic strength [56]. Analysis of these data showed that while both proteins were bound by heterogeneous interactions, that of lysozyme had more hydrophobic character than bovine serum albumin. These results suggest that net properties of the protein, such as the electrostatic and hydrophobic character, do not necessarily reflect the types of interactions that the protein have with immobilized dyes. In the second study, differences in the absorbance spectra of Cibacron Blue F-3GA under ionic or apolar conditions were observed. Thus it was proposed that the absorbance spectra could be used

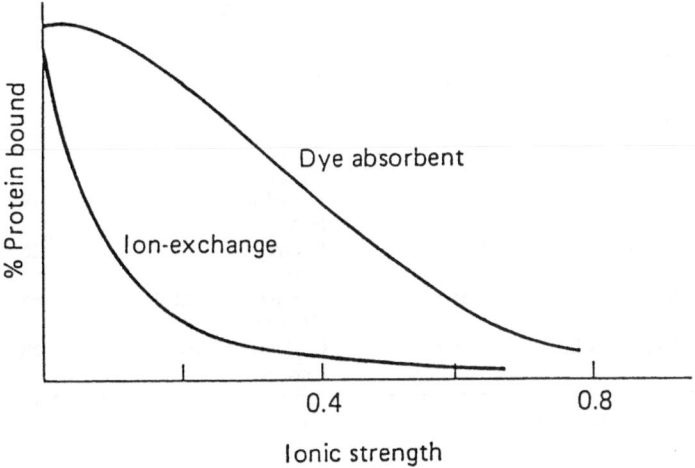

Fig 4. Effect of ionic strength on amount of protein (from a crude mixture) bound to dye adsorbent, compared with simple cation exchanger [34]

to determine the character of protein-dye interactions [57]. By this technique, it was found that a chicken liver dihydrofolate reductase was bound by electrostatic interactions, while a dihydrofolate reductase derived from *Lactobacillus casei* was bound by hydrophobic interactions.

These results merely serve to remind us of the complex nature of protein-reactive dye interactions. Not only can the reactive dyes participate in many types of interactions, a protein may contain several potential binding sites on its surface. Since the three dimensional structure of few proteins has been determined [58], it is unlikely that rational adsorbent design will be a viable option in many cases very soon. For these cases, empirically determined adsorbents based on merely vaguely understood interactions must suffice.

3.3 Metal Ion-Promoted Interactions

The adsorption of many proteins on immobilized triazine dyes are substantially enhanced in the presence of divalent metal ions. The adsorption of carboxypeptidase G_2 on Sepharose-bound Procion Red HE-8BN is one example of this phenomenon [59]. In a 100 mM Tris–HCl buffer at pH 7.3, only 15% of the total enzyme applied to the Procion Red HE-8BN column was adsorbed. However, when 0.2 mM Zn^{2+} was added to the enzyme sample being applied to the column under otherwise identical conditions, the enzyme was qualitatively adsorbed. A few other examples where metal ions have been observed to play an important role are listed in Table 2. Systematic studies of this phenomenon have shown that divalent metals, especially those in the first row transition series, are the most effective [60, 61]. As seen in Table 2, cases of metal-promoted binding of both metalloenzymes and proteins which contain no bound metal have been

Table 2. Examples of metal ion-promoted protein binding by immobilized reactive dyes

Protein	Source	Constitutive or cofactor metal ion	Reactive dye	Promoting metal ion	Ref.
Carboxypeptidase G_2	*Pseudomonas* sp.	Zn^{2+}	P. Red HE-8BN	Zn^{2+}	63
Alkaline phosphatase	Calf intestine	Zn^{2+}	P. Yellow H-A	Zn^{2+}	59
Hexokinase	*Saccharomyces cerevisiae*	Mg^{2+}	P. Green H-4G	Mg^{2+}	28
Tyrosinase	Mushroom	Cu^{2+}	P. Blue HE-RD	Zn^{2+}	64
Ovalbumin	Hen egg white	None	C. Blue F-3GA	Al^{3+}	59

Reactive Dye: P. = Procion; C. = Cibacron

observed, and the most effective metal ion does not always correspond to the constitutive one in the metalloenzymes [62]. Therefore, it does not seem likely that specific metal binding by the protein is responsible for the observed behavior.

Difference spectra studies have suggested that enhanced protein affinity results from the formation of highly specific ternary complexes involving the protein, dye and metal ion, rather than a simpler binary complex of just the dye and metal ion [60]. Application of these findings is relatively straight forward; in the case of hexokinase adsorption on Procion Green H-4G, the enzyme is adsorbed in the presence of Zn^{2+} and eluted by its omission in the eluant buffer [28]. In other cases, elution has been effected by inclusion of a chelating agent in the eluant [62]. Technical considerations in the application of divalent cations include solubility of the divalent cation in the presence of buffer salts as well as chelation of the metal ions by buffer salts [33]. An excellent review of the work completed in this field has recently been published [63].

4 Principles of Dye-Ligand Chromatography

Many classes of proteins have been purified by the technique of dye-ligand chromatography, using a wide variety of matrices and operating conditions. Scawen and Atkinson [7] have compiled an extensive list of recent reports of the application of dye-ligand chromatography for protein purification at the laboratory scale. This list demonstrates that, as in any adsorptive technique used for protein purification, the performance of a dye-ligand chromatography process relies on the proper choice of operating conditions. The first objective is to design an adsorbent which exhibits both high specificity and capacity for the desired protein. Buffer conditions need to be identified which promote selective binding of the protein in the adsorption stage, and selective desorption in the elution stage. Also the fixed-bed operating parameters such as flowrate, protein loading, and column geometry must be optimized to yield the most efficient process to meet the purification objectives.

4.1 Adsorbent Properties

The primary objectives in the development of a dye-ligand adsorbent is to produce a matrix which exhibits both high specificity and capacity for the desired protein. Several factors can affect the adsorbents specificity and capacity, including the matrix type, ligand type and concentration, as well as the use of a spacer arm. However, there are other important adsorbent properties such as physical and chemical stability of the matrix. These factors have important ramifications in the use of these adsorbents for large scale protein purification, especially in the case of therapeutic proteins where validation of the process is required.

4.1.1 Support Matrix

An ideal support matrix for a dye-ligand chromatography would exhibit the following properties: low non-specific protein adsorption, high binding capacity, chemical and biological stability, and physical incompressibility [5, 34, 48]. The requirement of low non-specific adsorption dictates that the matrix must be hydrophilic and contain few ionizable groups with which proteins can interact. In order to obtain high binding capacities, matrices have been chosen which have both very large pores, to facilitate the diffusion of proteins to internal binding sites, and a sufficient number of hydroxyl groups to allow adequate reactive dye immobilization. Furthermore, the matrix must be chemically stable in order to withstand the harsh conditions employed in cleaning the sterilization cycles [65]. If the adsorbent is to be used in a column, the matrix must be available in beaded form, and must exhibit adequate physical stability to be able to withstand the pressure required to force liquid through the bed.

These are fairly exacting requirements, and few matrices are available which satisfy all of them. Triazine dyes have been immobilized on a wide variety of matrices in the search for the optimal adsorbent. These include agarose (natural or cross-linked) [30, 33, 43, 44, 66–69], cross-linked dextran, [17, 27, 70, 71], beaded cellulose [72–75] polyacrylamide [76], Sepharcyl [13], glass [77], spheron [13, 78], and agarose-polyacrylamide copolymers [13]. Although studies comparing many of these matrices have indicated that cross-linked dextran matrices may allow more efficient ligand utilization [13], cross-linked agarose has remained the most popular matrix due to its relatively high incompressibility and open pore structure [79].

The main drawback of these agarose-based adsorbents is their unsuitability for use at high flowrates [71]. The large particle size (100 μm) and slow diffusion of proteins in the agarose matrix limit the resolution which can be achieved in columns of these adsorbents. Additionally, the agarose particles are compressible and consequently cannot withstand the pressures associated with high flowrates. In response to these limitations, the trend in recent years has been toward the use of incompressible matrices having smaller particle diameter. Examples of such matrices include Fractogel (Merck), TSK HW (Tosoh), Dynospheres (Dyno Particles), Superose (Pharmacia), as well as several inorganic matrices such as modified silicas and silicas coated with hydrophilic polymers. Recent studies have shown that dye-ligand adsorbents based on these matrices may be useful in protein purification processes at high flowrates [71, 161, 162], consequently, these newer matrices may begin to challenge agarose as the matrix of choice due to their excellent physical properties. Thus, at this time none of these matrices can claim to be universally superior; therefore, the choice of support is, by and large, still an empirical one which must reflect the needs of a particular application.

4.1.2 Ligand Selection

In the past ten years it has been recognized that almost all reactive dyes are able to interact with proteins, and that any one of these dyes may potentially act as a selective ligand for a particular protein. Due to both the unknown three-dimensional structure of most proteins (and in many cases the dyes themselves), as well as the seemingly random affinities which have observed between dyes and proteins, immobilized dyes have been termed "empirical" adsorbents [6]. Several methods have been suggested for identifying those reactive dyes which will be useful ligands for purification of a desired protein [27, 28, 31, 80]. The most common practice is to screen many small test columns of different immobilized dyes side-by-side for their ability to bind the protein of interest. Since the dye is already immobilized in these screening studies, this technique is more reliable than those methods which employ the free reactive dye, such as kinetic enzyme inhibition [27, 28]. Measurement of the unbound total protein and desired protein in the eluates from these small columns allows the identification of those dyes which selectively adsorb the desired protein [31]. Screening kits of this type are commercially available from a number of manufacturers including Amicon, Sigma, and Affinity Chromatography.

Screening 75 or more dye-ligand columns is a daunting task, therefore it is not surprising that several methods have been suggested which aim to reduce the amount of time and effort involved in this stage [33, 67, 68]. The most interesting of these ideas is the strategy proposed by Scopes in which the dyes are classified into five groups on the basis of their protein binding ability [33]. In this procedure a first round of screening, in which the objective is to identify the groups which are most likely to contain suitable ligands, involves only a few representatives from each of the five groups. This stage is followed by a second round of screening involving all of the dyes from the selected groups. By this two-stage procedure, almost half of the dyes are eliminated from the screening procedure. While this strategy is very elegant, there remains the drawback that a particularly useful dye may be overlooked if it belongs to one of the untested groups [68].

The other methods proposed for reducing the work involved in adsorbent screening, each rely on automation of the screening procedure, by either "home-made" or commercial devices [67, 68]. One interesting solution to this screening problem was the adaptation of a 96-well transplate cartridge for the purpose of adsorbent screening [68]. This method is particularly appealing because the eluates from the screening process can be collected in microtitre plates and assayed automatically for enzyme and protein by standard methods previously developed for ELISA techniques.

4.1.3 Spacer Arm

In affinity chromatography the ligand is usually coupled to the support matrix via a spacer arm molecule primarily to minimize steric hindrances on the protein-ligand interaction [12]. Spacer arms, however, are not commonly employed in the attachment of reactive dyes because of the ease of direct immobilization via the dye's reactive group. In a few cases, however, the use of a spacer arm has been found to enhance the immobilized dye's selectivity [81–83]. This result is most likely due to either reduced steric hindrance, as in the case of alcohol dehydrogenase adsorption to immobilized Cibacron Blue F-3GA analogues [55], or protein interactions with the spacer molecule itself, as recently reported for lactate dehydrogenase adsorption to spacer-mediated Cibacron Blue F-3GA [75]. Factors other than specificity, however, must also be considered regarding the use of spacer arms. Preparation of adsorbents for employing spacer arms requires activation of the matrix by treatment with toxic and hazardous chemicals [5, 12, 34, 45, 84], whereas direct immobilization avoids these hazards. In addition, spacer-mediated dyes are more prone to leakage than directly attached dyes [10–12]. These considerations are particularly important if the adsorbent is to be used in the production of a therapeutic protein, where the use of toxic chemicals and the leakage of ligand may become key issues in process validation. As a result, spacer arms are rarely used in attaching reactive dyes to the support matrix.

4.1.4 Immobilized Dye Concentration

The immobilized reactive dye concentration can have pronounced effects on the performance of a dye-ligand adsorption process, as binding capacity, affinity and selectivity have been found to vary as the ligand concentration is altered [19, 27, 85]. Early studies demonstrated that highly substituted adsorbents were generally able to bind more protein [13, 19], however, in a few cases elution of the desired protein was found to be more difficult in the highly substituted matrices [27, 85]. Chromatographic studies involving lactate dehydrogenase and alcohol dehydrogenase adsorption by immobilized Cibacron Blue have indicated that the increase in the apparent binding affinity may be due to multivalent interactions on highly substituted adsorbents [86–88]. Two recent studies have reported the effects of ligand concentration on equilibrium protein adsorption [44, 89]. Both found that the apparent dissociation constant decreased with increasing ligand concentration, and that binding capacity was proportional to ligand concentration at lower substitution levels. The deviation from proportionality observed for binding capacity in gels with high ligand concentration was found to be more pronounced for larger proteins, indicating that steric factors may be responsible for this behavior [44]. Dye concentration was also found to significantly affect the kinetics of fixed bed adsorption process, as forward rate constants were observed to decrease as ligand concentration was increased [44]. All of these results indicate that dye concentration is a very important factor in protein adsorption by immobilized dyes, and it should not be overlooked when developing a dye-ligand process. While these fundamental studies have begun to shed some light on the effects of ligand concentration on single component protein adsorption, the effects of this factor on protein adsorption from a crude mixture are still largely unknown and must be determined experimentally case by case.

4.1.5 Dye Leakage

In recent years the number of therapeutic proteins being licensed for production has grown dramatically. One side-effect of this development is the growing concern over leakage of chromatographic ligands into these products and their toxicity or pyrogenic effects [69, 90, 91]. In this regard, FDA rules mandate that no trace of the ligand may be present in material intended for human use [6]. The leakage of reactive dyes is less prominent than leakage from other affinity systems because the linkage is in general stronger [10–12], however, since the toxicity of reactive dyes in vivo has not been fully studied, the limit of safe exposure cannot be determined. These toxicological studies may very well determine the future application of dye-ligand chromatography in the production of therapeutic proteins [90].

The leakage rate of reactive dyes from chromatographic matrices is a function of pH, ionic strength and temperature [5]. Under conditions normally used in protein purification protocols, dye leakage is expected to be minimal because of the inherent stability of the adsorbents at these mild conditions. Since the amount of dye leached from these columns is very small, traditional spectroscopic measurements are not effective in dye leakage studies; therefore more sensitive methods must be used. Pearson and Lowe studied the leakage of Cibacron Blue from agarose by an ELISA technique, and found that leakage decreased from 0.65% in the first week to 0.05% by the fifth week [92]. In

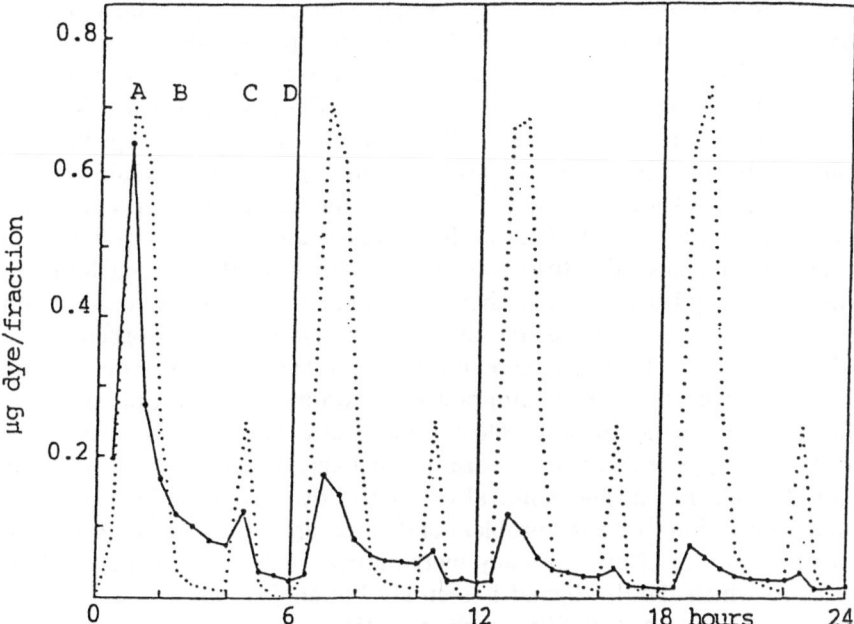

Fig 5. Leakage of dye from a ^{14}C Procion Blue HB-Sepharose column over four consecutive elution cycles: A = Fraction IV load; B = Wash; C = Albumin Desorption; D = Equilibration; *Solid line* = leakage of ^{14}C dye; *Dotted line* = optical density of eluate [14]

another study [14], (as shown in Fig. 5) the leakage of a C-labelled Procion Blue H-B dye from an agarose matrix was monitored throughout the adsorption process in an automated system for albumin purification. These results showed that after 18 months of use, 1.4% of the ligand had been leached from the column. Based on these results, each 25 g dosage of albumin produced in this system would be expected to contain approximately 10 µg of leached dye [90]. While this dose showed no apparent toxicity, it is clear that the effects must be studied.

Recent studies have reported that dye leakage from chromatographic supports was primarily due to matrix degradation rather than the disruption of the dye-matrix linkage [69, 93]. Thus it is clear that toxicological studies must involve not only the reactive dyes, but also the leached dye-matrix conjugates must be investigated. This was the subject of a recent study involving both Cibacron Blue F-3GA and leached agarose-Cibacron Blue F-3GA conjugates [69]. This study found no inhibition of the in vitro growth of human cells nor any significant increase in the number of polyploidic cells when cells were grown in the presence of the free Cibacron Blue F-3GA [69]. However, cells grown in the presence of the leached agarose-Cibacron Blue conjugate showed both inhibited growth and significantly higher frequencies of polyploidic cells [69]. Hence, it appears that the conjugated dye, rather than the free dye, poses the greater health risk. It is clear that much more investigation is needed in this subject.

These results suggest that removal of leached dyes from the finished product is very important if the product is intended for therapeutic use. The separation of the leached dye from the protein product may prove to be a relatively simple task, since the dye has much different properties than most proteins. Since at neutral values of pH, the dye carries multiple negative charges, passage through an anion-exchange column with a reasonably small-pore matrix should provide an efficient method for dye removal without significant loss of product [6]. Another possibility is the use of size exclusion chromatography or ultrafiltration since the dye is much smaller than most proteins; however, this approach should be used cautiously because the leached dye is linked to a polysaccharide chain which could be large. It is possible that leached dye may be adsorbed to the eluted proteins: in such a case, the removal of the dye is likely to be much more difficult. For dye-ligand chromatography to become a viable option in the purification of these therapeutic proteins, more study is needed in all these areas associated with dye-leakage: mechanism of leakage, detection of leached dye, toxicity of leached dyes, and removal of leached dyes from the final product. Another approach to solve the problem of dye leakage which has recently been proposed is the use of an extremely inert matrix, such as a perfluorocarbon polymer, to which the dye-ligand is securely bound [91, 94, 95]. The use of polyvinyl alcohol-coated perfluorocarbon matrix as a support for dye-ligand chromatography has recently been studied, and the results indicate that this matrix performed comparably to agarose-based adsorbents [94].

4.2 Buffer Conditions

Protein adsorption on dye-ligand adsorbents is very complex due to the mixed electrostatic/hydrophobic character of the protein-dye interactions. Consequently, in contrast to the cases of ion exchange and hydrophobic interaction chromatography, there are no reliable guidelines for adjusting buffer pH, ionic strength, or temperature to modify protein adsorption. Therefore, the development of a dye-ligand process typically involves experimental studies which are designed to identify suitable conditions. Due to the wide variety of compounds which have been shown to affect protein adsorption, including metal ions, detergents, nucleotide cofactors, this process of buffer optimization can become very involved. The following sections will detail how these various parameters have been manipulated in order to promote selective adsorption or elution of proteins, which is essential for successful operation of dye-ligand chromatography process.

4.2.1 Adsorption Stage

The primary objective of the adsorption stage of any chromatography process is to adsorb the desired protein quantitatively, while at the same time adsorbing as little of the contaminating protein as possible. As described in the preceding sections, the choice of ligand and its concentration in the adsorbent have a great impact on the adsorbent's selectivity. Buffer effects, however, are equally important, as a change in pH or ionic strength, or the inclusion of a nucleotide cofactor or metal ion can significantly alter the adsorption of either the desired protein or the contaminants. Thus optimization of buffer conditions used in the adsorption stage is an important step in the development of a dye-ligand process [32, 96, 97].

As mentioned before, protein adsorption from various crude preparations on immobilized reactive dyes generally decreases as either the pH or ionic strength of the buffer is raised. Similar results have also been found in protein adsorption studies involving pure proteins [56], but the extent of these effects was found to be protein-dependent. For example, raising the pH from 5.5 to 7.0 caused a 50% reduction in the adsorption of bovine serum albumin by immobilized Cibacron Blue F-3GA, while that of lysozyme was only slightly affected [56]. Differences in binding behavior between proteins form the basis for manipulation of adsorbent specificity by adjusting the buffer conditions, and as such, have been exploited for the purification of a number of proteins [32, 63, 85, 97–100]. An example of this type of manipulation is the isolation of diaphorase from *Bacillus stearothermophilus*, which is bound by immobilized Cibacron Blue F-3GA at high ionic strengths. Since very little contaminating protein adsorbs under these conditions, diaphorase could be purified to homogeneity in this single chromatographic step [100].

Temperature can also affect protein adsorption on immobilized dyes, and like pH and ionic strength, these effects vary from case to case depending on the types of interactions involved [48]. Operating temperatures are often determined by the stability of the desired protein, and usually fall in the range from 0 to 25 °C. However in cases where the protein is thermostable, alteration of this parameter to optimize specificity of adsorption should also be examined.

Other additives, including phosphate ions, divalent metal ions, buffer salts, chelating agents, and detergents have been observed to modify the affinity of dye-protein interactions [33, 63, 101]. A number of nucleotide-dependent enzymes exhibit decreased affinity for immobilized reactive dyes in the presence of phosphates, possibly as a result of competition between phosphate ions and charged sulfonate groups on the dye for the phosphate-binding sites on the enzyme [6, 33]. Conversely, it has recently been shown that phosphate may also participate in other types of protein-dye interactions, as demonstrated by prealbumin adsorption on immobilized Remazol Yellow GGL which requires the presence of phosphate or another anion for effective binding [102]. Divalent metal ions, as discussed previously, also promote protein adsorption through the formation of highly specific ternary complexes [62, 64]. As a result, protein adsorption from crude preparations generally increases when divalent metals, such as Mg^{2+} or Mn^{2+} are included in the adsorption buffer [33]. In cases where metal ions are involved in protein binding, some biological buffers including histidine, imidazole and, to a lesser extent, Tris have been found to interfere with protein adsorption due to chelation of the metal ions [59, 63, 64]. Thus the effects of these such buffer salts should be investigated in all cases where metal-promoted protein adsorption is suspected.

4.2.2 Elution Stage

A number of methods for promoting selective protein adsorption were discussed in the last section; however, since entirely specific protein adsorption is rarely achieved, techniques for selective elution of the desired protein must be developed. The objective of these techniques, whether they be specific or non-specific, is to completely elute the desired protein, at the same time, minimizing the amount of contaminating protein which is co-eluted.

Non-specific methods for protein elution almost always involve increasing the pH or ionic strength in the elution buffer [33]. These methods have been widely employed for protein elution from dye-ligand adsorbents in both stepwise and gradient techniques [28, 46, 71, 76, 85, 98, 103]. Electrostatic interactions between the protein and immobilized dye will be reduced under these conditions, thereby effecting the elution of proteins bound by this type of interaction [34]. Proteins which are bound by other types of interactions may not be released, as demonstrated by human interferon bound to immobilized Cibacron Blue F-3GA [104]. In this case, interferon could not be eluted by 1 M sodium chloride [104]; however, an elution buffer which consisted of 50% ethylene glycol affected complete elution [105], thereby indicating the hydrophobic nature of fibroblast interferon adsorption.

In cases where metal ions are involved in binding, it has been sometimes found that simply omitting the metal ion from the elution buffer can cause desorption [28]. However, in cases where the ternary protein-metal-dye complex is stronger, chelating agents such as EDTA must be added to the elution buffer in order to disrupt the complex [62, 63]. Along similar lines, in this example it was found that by switching from a Tris–HCl buffer system to Goods buffers, Hepes or Mops, recovery of carboxypeptidase G_2 from Procion Red H-8BN dropped significantly, probably as a result of chelation by the buffer salts [63].

Since the interaction of many proteins with immobilized dyes occurs at the protein's active site, affinity elution techniques have proven useful in a great number of cases [31, 32, 97, 105–107]. This technique can be very effective because the affinity eluant will interact specifically with the desired protein and cause it to desorb, while leaving the contaminating protein bound to the column [34, 52]. Nucleotide cofactors, such as $NADP^+$ [27, 31, 105, 106], NADH [100], ATP [50, 107, 108], and AMP [105] have been used to elute dehydrogenases and other nucleotide-dependent enzymes from immobilized Cibacron Blue F-3GA and other reactive dyes. In other cases, enzymes have been eluted by inclusion of one or more substrates in the elution buffer [29, 32, 97]. Considering that the reactive dyes are not "specific" ligands, affinity elution has found surprisingly wide application in dye-ligand chromatography. The success of affinity elution with these adsorbents may arise from the following facts: (1) many enzymes interact with immobilized reactive dyes by cation exchange at their active site [6, 28], and (2) most biological ligands are negatively charged [34]. Thus, many cases of affinity elution may result from competition at the protein active site between the biospecific ligand and the immobilized reactive dye [34, 52].

4.3 Operational Parameters

Fixed bed adsorption offers a number of advantages over stirred batch adsorption systems, including increased capacity and specificity, as well as decreased adsorbent handling. Therefore, except for cases where the process fluid contains particulate matter, the fixed bed configuration will usually be employed. Several factors must be considered in order to ensure the successful operation of these column systems, including liquid flowrate, column loading and column geometry. Not coincidentally, these comprise the primary parameters which must be considered in the translation of an adsorption process from laboratory to production scale [1].

The selection of the liquid flowrate in a fixed bed adsorption process is always a compromise between high resolution, obtained at low flowrates, and practical limitations which are determined by throughput requirements, column size and other factors [33]. Resolution in an adsorption column is related to the sharpness of the breakthrough curve and will generally be higher at low flowrates than at higher ones. Protein diffusion and adsorption in porous adsorbent particles are relatively slow, and at typical flowrates equilibrium is not approached [33]. Thus at low flowrates, the relative rates of diffusion and adsorption are increased, and result in sharper breakthrough curves. As demonstrated in a recent study involving protein adsorption by cation exchange [109], the apparent, or dynamic, binding capacity of a dye-ligand column decreases as the flowrate through the column is increased, simply as a result of reduced resolution [85]. Practical limitations, however, require the use of flowrates which are relatively high and consequently reduce the resolution of the column. In cross-linked agarose based adsorbents, such as dye-substituted Sepharose particles, superficial velocities of 20–30 cm h^{-1} have been suggested for adsorption columns at room temperature, while velocities of 10–15 cm h^{-1} are recommended for processes carried out at 4 °C due to decreased protein diffusivity at these temperatures [33]. Care should be exercised in applying these suggested flowrates, as the resulting pressure drop across the column may well exceed the limits of these soft matrices. As mentioned earlier, new adsorbent matrices have been introduced which allow the use of higher flowrates. These particles are both smaller, which allows increased resolution at high flowrates, and stronger in order to accommodate the increased pressures which accompany such high flowrates. Several of these matrices have been shown to be

suitable for use in dye-ligand chromatography columns operating at high
flowrates [71].

The amount of protein loaded onto a column can also have pronounced
effects on adsorption specificity. If the interaction between the desired protein is
relatively strong, then loading the column until that protein begins to breakthrough may be desirable. In this case underloading the column may result in
weakly bound contaminants co-adsorbing on the adsorbent, which in overload
operation would be displaced by the desired protein [33]. Examples of protein
overloading on dye columns include the purification of yeast AMP kinase and
Alcaligenes eutrophus lactate dehydrogenase [33]. If the desired protein interacts weakly, however, underloading may be necessary to maintain the specificity
of the adsorption stage [48].

In laboratory adsorption columns, the geometry of the fixed bed is rarely
considered a critical parameter [45], however, in large-scale processes column
geometry can have significant effects. Laboratory columns generally have a
reasonably high height-to-diameter ratio, or aspect ratio, often greater than 10,
whereas process scale columns are usually short and squat, with aspect ratios
generally than 3. Short, squat columns are used in large-scale because the soft
adsorbent matrices cannot withstand the hydrostatic pressure which occurs in
tall columns, yet large column volumes are required. The long thin geometry of
the laboratory column allows more efficient use of the adsorbent capacity than a
short, squat column operated at the same superficial velocity, simply due to the
increased contact time in the long column. Effects of column geometry are also
apparent during elution, as it is commonly observed that the desired protein is
eluted in a smaller volume from a long thin column than from a short column of
an equal volume [85]. It is likely that some of these differences in scale may be
alleviated by the use of new, rigid matrices, by allowing process scale columns to
have similar geometry to those at the laboratory scale.

5 Quantitative Analysis in Dye-Ligand Chromatography

Interest in the modelling and simulation of protein adsorption processes has
grown in recent years, out of the need for reliable methods for the design and
scale-up of chromatographic protein purifications. The models being developed
account for adsorption kinetics and equilibria as well as various mass transfer
effects which occur in protein adsorption [110, 111]. As a result, research
concerning the quantitative aspects of dye-ligand and other affinity adsorption
processes have become an area of intense activity [44, 56, 112–120]. Several
methods which have been used to determine the kinetic and equilibrium
parameters for protein adsorption will be presented in the following sections.

5.1 Equilibrium Parameters

The most fundamental characterization of the interaction between an immobilized ligand and a protein is that of equilibrium adsorption. Equilibrium isotherms give information regarding both the affinity of the interaction and the binding capacity of the adsorbent. The most common isotherm model is the Langmuir isotherm, which may be derived from the assumption a ideal, homogeneous second order interaction between the immobilized ligand and protein [112]. As seen in Table 3, this model contains two parameters: the maximum binding capacity (Q_m) and the dissociation constant (K_d). The effects of these parameters on the performance of batch [121] and fixed bed adsorption processes [112] have been discussed previously. Table 3 also shows the form of the Freundlich isotherm, another commonly used model for adsorption equilibrium. Several methods have been used to determine isotherm parameters for protein adsorption in dye-ligand adsorbents, and these methods are described in the sections below.

5.1.1 Kinetic Enzyme Inhibition

Equilibrium parameters such as dissociation constants, binding stoichiometry, and inhibition constants are commonly associated with enzyme kinetics. Since many reliable methods have been established for the characterization of enzyme kinetics [122], it has been proposed that the equilibrium constants for the protein-immobilized ligand (i.e. reactive dye or other affinity ligand) interactions may be conveniently determined by kinetic enzyme inhibition studies [124]. Such inhibition studies have been carried out for a number of enzymes with triazine dyes [28, 29, 47, 51, 123]. In many of these cases, this approach has been proven an effective means of identifying reactive dyes which may serve as suitable ligands, as well as a means of identifying buffer conditions which promote interactions between the enzyme and reactive dye [28, 29]. However, some caution is advised in the application of these results to chromatographic

Table 3. Models for equilibrium adsorption

Langmuir Isotherm

$$C_s^* = \frac{Q_m C^*}{K_d + C^*}$$

Freundlich Isotherm

$$C_s^* = k(C^*)^{1/n}$$

Notation is given in the List of Symbols

systems. Immobilization of the reactive dye may significantly alter the protein-dye interaction due to differences in the local environment inside the matrix and steric constraints on both the immobilized dye and the protein. This point is illustrated well by the results of Burton and co-workers [54, 55], who found that slight modifications of the terminal ring of Procion blue H-B led to significant enhancement of the binding affinity of the dye for alcohol dehydrogenase, as determined by kinetic inhibition studies. However, immobilization of the react-ive dye analogues by direct attachment via the triazine ring led to a great reduction in the affinity of the dye-protein interaction. Thus it is seen that equilibrium parameters determined by homogeneous methods using the free dye may not be valid for the interaction of the protein with the immobilized dye, and therefore may not be suitable for use in models developed to describe the protein adsorption process.

5.1.2 Chromatographic Methods

In order to avoid these potential problems associated with the use of free ligands, chromatographic techniques have been developed which allow the characterization of protein interactions with immobilized ligands. Various models have been derived based on classical enzyme kinetics to describe these systems, including zonal or frontal analysis and monovalent or multivalent interactions [124–127]. Experimental procedures involve introduction of either a pulse or a step in protein concentration to a fixed bed of adsorbent, often competitive inhibitors are included in the mobile phase. Several studies of this type have been conducted with dye-ligand adsorbents typically using mobile phase competitors such as NAD^+ or NADH [86–88, 128, 129], although a recent study reported the use of a Cibacron Blue F-3GA-dextran conjugate as a competitive inhibitor [73]. These chromatographic methods have been applied to the case of lactate dehydrogenase adsorption on immobilized Cibacron Blue F-3GA, and have indicated that this tetrameric protein is bound by a single ligand at its active site [129]. Similar studies with horse liver alcohol dehydro-genase have shown that adsorption of this dimeric enzyme is monovalent in absorbents with low immobilized dye concentration, and multivalent in highly substituted adsorbents [86, 88]. Chromatographic methods are also useful because the adsorbent capacity, or effective ligand concentration, may be determined by either frontal or zonal operation of the column [86, 87, 129]. Thus, these methods can be a very effective means of characterizing protein-ligand interactions in cases where the adsorption is specific (i.e. binding occurs at the active site), however, they have not been extended to cases where non-specific binding interferes or dominates the interactions. Nevertheless, chro-matographic methods can be used to acquire reasonable estimates of the adsorbent's binding capacity and affinity for the protein, and these parameters are more likely to be valid than those determined by the kinetic enzyme inhibition studies described in the previous section.

5.1.3 Batch Adsorption Equilibrium Methods

The chromatographic procedures are best suited for the characterization of affinity adsorption where the ligand-protein interaction is specific, and therefore can be expected to obey classical enzyme kinetics. Protein adsorption by immobilized dyes, however, rarely conforms to these simplified assumptions due to the complexity of the interactions. Simple and direct measurements of equilibrium adsorption have been shown to be an effective means for obtaining equilibrium parameters for these nonideal cases [44, 56, 112]. Experimentally, this method is very simple: known quantities of protein are contacted batch-wise with adsorbent, and at equilibrium samples are taken to determine the protein concentrations in both the solution and, by difference, in the adsorbent [112]. These experiments are not subject to some of the errors associated with column systems such as axial dispersion, non-uniform packing, and liquid channelling. Furthermore, since the equilibrium data are obtained experimentally, isotherm models may be evaluated by direct comparison with the data, rather than indirect methods involving idealized chromatography models [44, 56, 130–132].

Using this batch adsorption procedure, the equilibrium adsorption of a number of proteins on dye-ligand adsorbents has been characterized [44, 56, 112, 118, 119, 130–132]. In many cases the Langmuir isotherm has been shown to yield an adequate fit to the experimental data [44, 112, 113, 118, 119, 130–132]; however, in a few cases the empirical Freundlich isotherm fit the data much better [56, 132]. This result underscores the fact that many proteins are non-specifically bound by the immobilized dyes. In many cases, protein adsorption may involve heterogeneous, as well as multivalent, interactions [56]. Thus, batch adsorption methods may give more insight into the character of protein binding because it allows a more objective evaluation of the adsorption equilibrium than is possible using the chromatographic methods.

5.1.4 Typical Values of Equilibrium Parameters

The isotherm parameters for adsorption equilibrium of a number of proteins have been reported, however as discussed in a previous section, these parameters vary significantly under different conditions of pH, ionic strength, and buffer system [56, 87]. In addition, both the binding capacity and affinity vary as the immobilized dye concentration in the adsorbent is varied [44, 86, 89, 131]. Thus, reported values of isotherm parameters must be considered valid only under the conditions used in that study. However, much can be learned by comparing typical isotherm parameters for protein adsorption on immobilized dyes with that of other types of adsorbents under typical adsorption conditions. The binding capacity of dye-ligand adsorbents is generally observed to be higher than that of biospecific affinity adsorbents [45], and lower than that of a typical ion-exchange adsorbent, although some highly substituted dye-ligand adsorbents have been reported to exhibit capacities which are comparable to those of ion exchange (~ 100 mg ml^{-1}) [44]. Typical values of apparent dissociation constants obtained for protein adsorption on immobilized dyes usually fall between 10^{-6} and 10^{-7} M [44, 112, 118, 120]. These values are of similar magnitude to those found by chromatographic methods [86, 129]. Apparent dissociation constants for protein adsorption on ion exchange adsorbents have reported to be on the order of 10^{-6} M [115, 133], while that of biospecific [114] and immunoaffinity adsorbents has been reported to be 10^{-8} M [134]. Thus, these data show that the affinity exhibited by immobilized reactive dyes is intermediate between ion exchange and biospecific affinity adsorbents.

5.2 Adsorption Kinetics

The equilibrium parameters discussed above are very important because they determine the amount of protein which can be bound under a given set of conditions; however, of equal importance in the design of an adsorption process is the rate at which adsorption occurs. Adsorption in a fixed bed is dynamic and depends on both physical and chemical processes. The physical processes involve mass transport of the protein, which may include axial dispersion, film diffusion, and diffusion inside the adsorbent particle [116, 135]. The chemical processes include the adsorption and desorption steps as well as other non-ideal interactions such as protein spreading [135]. These various processes combine to make the study of protein adsorption kinetics very complex, even for the ideal case of a second-order reversible interaction between the protein and immobilized ligand. Further complications arise for the case of immobilized dyes due to the "non-ideal" interactions which may be multivalent and heterogeneous. Also, since immobilized dyes usually interact differently than free dyes, it is difficult to isolate the interaction kinetics from the mass transport processes. As a result, mathematical rate models, which are based on material balances and rate equations, must be employed to evaluate rate parameters. Perhaps the simplest rate model is the one presented by Thomas for ion exchange [136], where the interaction and mass transfer rates were lumped into a single rate equation. Chase [112] and others [131, 137] have demonstrated the use of this equation in predicting the behavior of dye-ligand columns based on experimentally determined parameters. Modifications of this simple model have been suggested to account for the effects of mass transport [138], and these methods have been applied to protein adsorption by immobilized dyes [44, 118]. These results suggest that intrinsic protein adsorption is rate-limiting in these systems. Other more rigorous models have been developed to describe kinetic protein adsorption in fixed-bed systems [110, 111]. An example of such a rate model for protein adsorption is shown in Table 4. Although these rigorous models have not been used to evaluate kinetic parameters for dye-ligand adsorption, their application in ion exchange [115, 139] and affinity protein adsorption [114] have suggested that slow intraparticle diffusion, rather than slow intrinsic kinetics, was rate-determining. It is clear that much more work is needed to clarify these

Table 4. Rate model for protein adsorption on a fixed-bed of porous adsorbent

---

Material Balance on Bulk Phase:

$$\frac{\partial C}{\partial t} + V\frac{\partial C}{\partial z} - D_L\frac{\partial^2 C}{\partial z^2} = -\frac{3k_F(1-\varepsilon)}{\varepsilon R}(C - C_p|_{r=R})$$

Material Balance on Stationary Phase:

$$\varepsilon_p\frac{\partial C_s}{\partial t} + \frac{\partial C_s}{\partial t} = \frac{\varepsilon_p D_p}{r^2}\frac{\partial}{\partial r}\left[r^2\frac{\partial C_p}{\partial r}\right]$$

Adsorption Rate Equation:

$$\frac{\partial C_s}{\partial_t} = k_1 C_p(Q_m - C_s) - k_2 C_s$$

Notation is given in the List of Symbols

contradictory results and identify the true rate-limiting steps in protein adsorption processes by dye-ligand and other adsorbents.

6 Large-Scale Dye-Ligand Chromatography

As discussed previously, the application of adsorption processes at production scale often requires careful attention to several factors which may be only minor considerations at the laboratory scale. These include physical and chemical stability of the adsorbent matrix, ease of ligand immobilization, ligand stability, the ability to regenerate and reuse the adsorbent, reproducibility, and process validation. An important factor in large-scale adsorption, which is rarely considered in laboratory research, is cost. While these factors have limited the application of affinity chromatography in large-scale protein purification, dye-ligand chromatography appears promising in this regard.

The major considerations in the application of dye-ligand chromatography at process-scale involve the physical and chemical stability of the immobilized dye adsorbents [65]. As discussed in the previous section, large columns are needed in these processes, yet the column height is usually limited to less than 100 cm by the compressibility of the matrix. Hence process scale columns tend to be short and squat, a geometry which has been shown to compromise both the adsorbent capacity and the effectiveness of the elution stage [85]. Due to these limitations, it is likely that newer, more rigid matrices will begin to replace the gels presently used in large-scale chromatographic processes [71].

The high cost of adsorbents and the labor involved in packing the adsorption columns require the regeneration and reuse of adsorption columns. Since the conditions employed in column cleaning and sterilization are harsh, the immobilized ligand and the matrix must both exhibit great chemical stability. One possible criterion, which has been suggested, is that the adsorbent must be able to withstand 1 N sodium hydroxide, a treatment often used for depyrogenation [65]. If the immobilized dye is slightly altered through contact with harsh buffers routinely used in column operation, however, these altered dyes often exhibit similar affinities for the desired proteins. Since immobilized reactive dyes and agarose are reasonably stable under these conditions, adsorbents based on these components should be suitable for large-scale processing [65].

There have been several reports of large-scale protein purification schemes using dye-ligand chromatography [14, 51, 85, 96, 100, 103, 140–145]. Most of these procedures have been developed for the purification of either nucleotide-dependent enzymes, such as dehydrogenases or kinases, or for blood-derived proteins. A list of some of the proteins for which large-scale purifications involving the use of immobilized reactive dyes is shown in Table 5. The role of dye-ligand chromatography in these processes has varied widely. For example, chromatography on immobilized Cibacron Blue F-3GA was used only as the final, polishing step in the purifications of two phosphotransferases from beef heart [140, 141]. In other applications, such as the purification of horse liver alcohol dehydrogenase [96] and *B. stearothermophilus* glycerokinase [85] and glucokinase, [107], dye-ligand chromatography was an important step in the multistep purification schemes. In the case of malate- and hydroxybutyrate dehydrogenases from *Rhodopseudomonas spheroides*, adsorption on immobilized Procion Red H-3B was employed as the first step after homogenization, indicating the high selectivity of the dye for these enzymes [7, 142]. The purification of diaphorase from *B. stearothermophilus* to homogeneity in a single chromatography step is an impressive display of the selectivity achievable by these immobilized reactive dyes [100]. In this scheme diaphorase was adsorbed selectively onto immobilized Cibacron Blue F-3GA at high ionic strength (2 M KCl), and eluted by a gradient of NADH. Lastly, one of the most encouraging applications of dye-ligand chromatography has been the full-scale production of human serum albumin by adsorption on immobilized Procion H-B [14]. A fully automated adsorption system, which can operate unattended through 30 adsorption cycles and is based on a 40 l column has been developed for this process [14]. These various examples demonstrate well the ability of dye-ligand chromatography to fulfill a variety of roles in the large-scale production of proteins.

Perhaps the most important question in the area of dye-ligand chromatography, however, is not concerned with its technical feasibility, rather it is whether these adsorbents will be licensed for use in the production of therapeutic proteins. Dye leakage has been reported to occur in both the adsorption and elution stages, therefore the leached dye will likely be present as a contaminant in the final product [14]. Currently, the long term physiological effects of these dyes are unknown [69, 90]. Dye leakage can be minimized by the

Table 5. Examples of large-scale protein purifications which employ dye-ligand chromatography

Protein	Source	Matrix	Ligand	Eluant	Ref.
Albumin	Human serum	CLA	P. Blue H-B	3 M NaCl or 20 mM Na Octonate	14
Alcohol dehydrogenase	Horse liver	A	C. Blue F-3GA	10 mM NAD^+ + pH 5.9	96
Carboxypeptidase G_2	Pseudomonas sp.	A	P. Red H-8BN	0.1 M EDTA + pH 7.3	63
Chymosin	Calf rennet	A	C. Blue F-3GA	0.85 M NaCl	98
Diaphorase	Bacillus stearothermophilus	A	C. Blue F-3GA	0–2 mM NADH	100
Enterotoxins A, B, and C_2	Staphylococcus aureus	CLA	P. Red HE-3B	60 mM phosphate, 150 mM phosphate or 35 mM KCl	103
Glucokinase	Bacillus stearothermophilus	A	P. Brown H-3R	5 mM ATP	107
Glycerokinase	Bacillus stearothermophilus	A	P. Blue MX-3G	5 mM Mg-ATP	85
Hyroxybutyrate dehydrogenase	Rhodopseudomonas spheroides	A	P. Red H-3B	1 M KCl	142
Lactate dehydrogenase	Human haemolysate	CLA	P. Blue HE-GN	5 mM AMP	105
Lactate dehydrogenase	Rabbit muscle	S	P. Blue MX-R	7 mM NADH	155
Malate dehydrogenase	Rhodopseudomonas spheroides	A	P. Red H-3B	1 M KCl + 2 mM NADH	142
6-Phosphogluconate dehydrogenase	Human haemolysate	CLA	P. Blue MX-4G P. Blue HE-GN	0–0.7 M KCl 2 mM hydrolysed NADP	105
Phosphoglycerate kinase	Saccharomyces cerevisiae	A	C. Blue F-3GA	0–1 M NaCl	143
ATP–AMP phosphotransferase	Beef heart	CLA	C. Blue F-3GA	0.4–2 M NaCl	141
ATP–AMP phosphotransferase	Saccharomyces cerevisiae	CLA	C. Blue F-3GA	0.5–3 M NaCl	144
GTP–AMP phosphotransferase	Beef heart	CLA	Blue Dextran	1 mM mGTP	140
Protein kinase	Porcine liver	A	Blue Dextran	0–3 mM ATP + 0.4 M NaCl	145

Matrix: A = agarose; CLA = cross-linked agarose; S = silica

complete removal of the unbound dye after immobilization. Since at large scales the batch-wise washing procedure [30] requires extremely large volumes of cleaning solutions, it has been suggested that more efficient cleaning may be achieved by packing the adsorbent into a column and washing the column [7]. As mentioned earlier, removal of the leached dye from the final product appears to be feasible, however, no systematic testing of such a procedure has been reported. These issues may well determine the fate of large-scale dye-ligand chromatography, and therefore need to be resolved.

7 Tandem Dye-Ligand Chromatography

One of the most interesting developments in dye-ligand chromatography was the introduction of tandem (or differential) dye-ligand chromatography [31]. As is evident from its name, this technique employs two immobilized dye columns, one which exhibits "negative selectivity" for the desired protein (i.e. high selectivity for contaminating protein) and a second which displays "positive specificity" (i.e. high selectivity for the desired protein). The underlying principle of this system, as shown in Fig. 6, is that contaminating protein is removed by the negative column before adsorption of the desired protein on to the positive column. In this system the crude protein mixture is fed to the negative column, where much of the contaminating protein is bound. The desired protein, however, does not adsorb on the negative column and appears in the eluate, which is fed directly to the positive column, where the desired protein is adsorbed. After this adsorption stage is complete, the two columns are disconnected and the second column is independently washed and eluted to selectively recover the desired protein.

Tandem dye chromatography has significant advantages over systems in which either a single positive column or two consecutive positive columns are employed. Inclusion of a negative column before a single positive column allows the removal of much of the contaminating protein, which may otherwise have been adsorbed on the positive column along with the desired protein. By this means, the amount of protein introduced to the second column is greatly reduced [34], and in effect its selectivity for the desired protein is increased. The advantages of tandem columns over consecutive positive columns include increased processing speed and recovery. In tandem operation, since the adsorption stage for both columns proceeds simultaneously and only the second column requires elution, the time required is much less than that for two positive columns where each must be adsorbed and eluted sequentially. While a reduction in processing time in general is attractive, rapid purification of labile proteins can be critical. For example, the characterization of the extremely labile enzyme, 6-phosphogluconate dehydratase from *Zymomonas mobilis* was made possible by rapid purification in a tandem dye-ligand chromatography process,

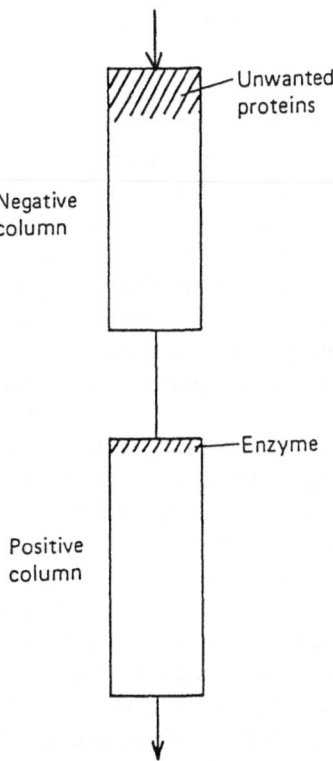

Fig 6. Principle of differential column chromatography [34]

whereas previous attempts using conventional techniques failed [97]. Further-more, tandem systems should allow higher recoveries than consecutive columns due to the fact that the desired protein is eluted and collected only once in this technique rather than twice. Lastly, another advantage will be realized if the process is to be implemented at production scale and requires validation. Validation costs increase with the number of chromatographic fractions to be analyzed, and these costs can represent 10–50% of process development costs for chromatographic processes [146]. Since only one set of fractions is generated in the tandem column system, validation costs should be much lower than a system employing two positive columns.

The development of a successful tandem dye-ligand chromatography pro-cess relies primarily on the identification of suitable negative and positive adsorbents. A good negative adsorbent should exhibit negligible adsorption of the desired protein, and both high affinity and capacity for contaminating proteins. A positive adsorbent, on the other hand, should exhibit high selectivity for the desired protein, adequate capacity, and quantitative elution at mild conditions. As before, identification of suitable adsorbents is accomplished through dye screening [31, 32], or by one of the strategies developed to reduce this task which were discussed previously [33, 67, 68]. These screening proced-ures result in the selection of a few "candidate" dyes, which must be subjected to a battery of testing in order to optimize dye concentrations and buffer conditions.

Tandem dye-ligand chromatography has become recognized as one of the more selective chromatographic methods available for protein purification [6]. This technique has been used to isolate a number of enzymes, including glucose-6-phosphate dehydrogenase from *Leuconostoc mesenteroides* [31], and several enzymes which are active in the Entner-Douderoff pathway in *Zymomonas mobilis* [32, 97, 106, 147]. A particularly interesting example is the simultaneous purification of glucokinase, fructokinase, and glucose-6-phosphate dehydrogenase from an extract of *Z. mobilis* by differential chromatography methods employing three tandem immobilized dye columns [106]. This method is very effective, as these enzymes have been purified to near homogeneity with overall recoveries ranging from 65 to 90%.

While the technique of tandem chromatography is particularly suited for use with immobilized reactive dyes due to the wide range of subtly different affinities exhibited by these adsorbents, it is by no means limited to these adsorbents. The principle of tandem chromatography, namely the application of a multiple column system in which contaminating protein is removed before the desired protein is selectively adsorbed, may prove useful with other classes of adsorbents as well. As discussed previously, the effectiveness of a positive adsorption column may be substantially increased by including a negative dye-ligand column upstream, and this may offer a simple method for improving the effectiveness of many existing protein adsorption schemes. Thus, the use of tandem chromatography may potentially spill over into the more traditional areas of ion exchange or even affinity chromatography.

8 High Performance Dye-Ligand Chromatography

The technique of high performance liquid chromatography (HPLC) has gained wide acceptance as an analytical tool and is attracting increasing interest as a potential method for preparative protein purification [148]. HPLC offers the advantages of rapid separation and high resolution, and has been successfully combined with affinity chromatography to yield a technique which offers high specificity as well [149]. This technique has been termed high performance liquid affinity chromatography (HPLAC), and its utility has been demonstrated in the resolution of a number of proteins, enzymes and isoenzymes using a variety of immobilized ligands [149–151]. Due to the wide range of proteins which interact with reactive dyes, several researchers have studied the application of these dyes as general ligands in HPLAC [152–162].

HPLC is characterized by its use of very small particles as well as the high pressures required to force liquid through these microparticulate columns at high flowrates. Thus, the matrices used in HPLC columns must be incompressible. Microparticulate silica has been used in the vast majority of HPLAC applications, mainly due to the lack of competing matrices [148, 158]. There are a number of drawbacks in the use of silica matrices in HPLAC [158]. The first drawback is the requirement for surface modification of silica by organosilanes [163] in order to create an inert, hydrophilic surface which is suitable for ligand immobilization. A second limitation is the lack of chemical stability exhibited by silica adsorbents in even mildly alkaline conditions. Since the solubility of silica is significant at pH as low as 8.0 [93], typical procedures [39] for reactive dye immobilization and adsorbent regeneration are not suitable with these adsorbents. These disadvantages, as well as limitations of silica pore size (less than 30 nm), have led to the development of a number of synthetic matrices suitable for use in HPLAC applications, including TSK G5000 PW (Toyo Soda) [158], Dynospheres XP-3507 (Dyno Particles) [158, 161], and Monospheres (Merck) [159]. The suitability of several of these hydrophilic, rigid matrices as well as silicas coated with similar hydrophilic polymers for use as supports in HPLAC purifications of enzymes has been investigated, and several were found to be suitable [158, 159, 161, 162]. A procedure has been described [152] for the immobilization of aminohexyl reactive dyes on modified silica; the resulting adsorbents contained 5.5–6.7 µmole dye per g silica. An alternative method of direct dye attachment was described by Clonis [158] yielded adsorbents which contained 3.0–8.5 µmole dye per g silica. Immobilization of reactive dyes on the synthetic matrices have been accomplished by methods which are similar to those used for agarose matrices [158], and immobilized dye concentrations obtained varied from 5–23 µmole dye per g dry gel.

Several experimental studies have demonstrated the resolution and purification which can be achieved by HPLAC [152–155, 158, 159, 161]. The resolving power of these adsorbents has been shown to be very high, especially when affinity elution techniques are employed. Using this method, dehydrogenase enzymes as well as kinase enzymes could be resolved by affinity elution techniques involving the use of cofactors and substrates [152]. While modified microporous silica has been used in the majority of these applications, [152–155, 160] other matrices have been investigated, including several synthetic matrices [158, 161], dextran-coated silica [71, 156, 157], cellulose-coated silica [162], and non-porous silica [159].

The use of HPLAC with immobilized reactive dyes at preparative scale has been reported [154, 155]. The operation of a 3.3 l column containing Procion Blue MX-R-Silica for the production of lactate dehydrogenase has been shown to yield 8.6-fold increase in specific activity with a recovery of 50% [155]. In a smaller preparative-scale system employing a similar adsorbent, a 5.1-fold increase in specific activity of lactate dehydrogenase with 80% recovery was achieved. The capacity of this adsorbent, as determined by frontal analysis, was determined to be 0.33 mg pure lactate dehydrogenase per g adsorbent [154]. The equilibrium adsorption of lactate dehydrogenase on two similar adsorbents has been recently characterized [160]. The measured binding capacity was found to vary with the concentration of immobilized dye and the purity of the protein mixture, with reported values ranging from 1.7–8.7 mg g^{-1} [160]. It is likely that these differences in the values for binding capacity simply reflect differences in the crude protein mixtures employed. It is clear from these examples that the HPLAC techniques based on immobilized reactive dyes show great potential on the preparative scale; however, more work is needed to evaluate the performance of these various new matrices and further characterize these adsorbents.

9 Non-Chromatographic Techniques Using Reactive Dyes

The reactive dyes have been shown to exhibit several advantages over other affinity ligands, such as stability, ease of immobilization, and high selectivity. These properties have led to the use of reactive dyes as affinity ligands in several non-chromatographic techniques for protein purification. These techniques include affinity partitioning in aqueous biphasic systems, affinity membrane separations, affinity cross-flow filtration, and affinity precipitation.

Protein partitioning in aqueous biphasic systems, such as polyethylene glycol-dextran, can be greatly enhanced by the attachment of an affinity ligand to one of the phase-forming polymers [164]. Reactive dyes have found wide application in these systems, and in many cases have been responsible for increasing the partition coefficient for a desired protein by several orders of magnitude [165]. Procedures have been described for immobilizing the reactive dyes on dextran [89] and polyethylene glycol [166], two of the most commonly employed phase-forming polymers. These enhanced partition coefficients can often be translated into effective protein extractions by adjusting the phase system conditions to minimize the partitioning of contaminating protein into the dye-ligand phase. Systems of this type have been developed for preparative-scale purification of several proteins [167–170]. Studies have also shown that affinity partition can be used effectively in unclarified homogenates, thereby eliminating the need for clarification as in chromatographic systems [169, 170]. Affinity partitioning is also attractive because of the ease in scaling up these purification procedures, thus this method shows great promise [171].

Affinity adsorption on membranes has been proposed as an alternative to the traditionally employed soft beaded gels, in order to alleviate the limitations of slow flowrates, adsorbent compressibility, and slow mass transfer which are associated with these gels [172, 173]. In these systems, the affinity ligand is immobilized on the surfaces of the membrane, thereby eliminating the slow diffusion step found in porous matrices and consequently allowing the use of higher flowrates [174]. Recent studies using membrane-bound reactive dyes have found that the capacities of membrane and agarose-based adsorbents were similar in magnitude in batch adsorption [132, 175]. However, under frontal loading conditions, a Cibacron Blue-membrane adsorbent exhibited a higher binding capacity than a similar agarose adsorbent, Furthermore, at very high flowrates the membrane showed almost no decrease in binding, while the agarose adsorbent was observed to drop significantly. These results demonstrate the advantages possible in affinity membrane separations. Large-scale applications of this technique will most likely involve the use of hollow-fiber configurations in order to achieve the high surface areas required [65, 176, 177].

Affinity cross-flow filtration is another technique which employs a membrane as a means of separating proteins. In this system an affinity ligand is immobilized on a soluble or insoluble macro-ligand which is too large to pass through the ultrafiltration membrane [178]. Using this configuration, the desired protein is bound to the macro-ligand and held in the ultrafiltration retentate, while all other contaminating proteins are washed through the ultrafiltration membrane. The desired protein may then be recovered by changing buffer conditions to effect elution and passing the mixture back through the ultrafiltration unit. Cibacron Blue F-3GA has been immobilized to a variety of supports and used in such systems to purify several dehydrogenase enzymes [179, 180]. Furthermore, theoretical models have been developed and tested using the model systems of human serum albumin and lysozyme adsorption on Cibacron Blue F-3GA-agarose [130, 131]. The successful implementation of this technique relies heavily on proper choice of the macro-ligand [181].

One last use of reactive dyes for protein purification is fractional affinity precipitation. This application differs from the other methods in that the reactive dye is not immobilized. Affinity precipitation involves the formation of an insoluble protein network by cross-linking with a multifunctional ligand. Reactive dyes and dye analogues have been shown to be effective precipitants due to the specific interactions of these dyes with several proteins [182–184]. Examples include the precipitation of plasminogen by the bifunctional dye Procion Red HE-3B (shown in Fig. 1) as well as precipitation of lactate dehydrogenase by a Procion Blue H-B derivative. A complete review of the use of reactive dyes has recently appeared [185].

10 Conclusions

In the last ten years, Cibacron Blue has persisted as the most widely used reactive dye for protein purification, although today it is widely recognized that others may serve equally well as affinity ligands. Due to the complex nature of the dye-protein interactions, which often involve both electrostatic and hydrophobic forces, an empirical approach is often pursued with these adsorbents. Screening procedures have been modified to facilitate the screening of 75 or more commercially available reactive dyes for the identification of suitable ligands. An alternative to dye screening is the rational design of biomimetic dyes via computer simulation of the dye-protein interactions. While at this time only a few dye-protein interactions have been characterized well enough to allow the application of this method, this technique is likely to become very important in years to come.

Buffer conditions and added compounds, such as divalent metals and nucleotide cofactors, play important roles in protein adsorption, and these factors have been exploited to maximize the purification of the desired protein. Adsorbent properties, including ligand concentration and matrix selection also have important effects on the performance of dye-ligand chromatography. Leakage of the reactive dyes into the final product remains to be a great consideration, as the toxicity of these leached dyes has not been fully characterized.

Further characterization of the equilibrium and kinetic processes involved in protein adsorption should lead to the development of more reliable design and scale-up procedures, which should in turn lead to wider application of this technology at the production-scale. Thus, the application of reactive dyes in protein purification appears to be a promising technology.

11 References

1. Janson J-C, Hedman P (1982) In: Fiechter A (ed) Advances in biochemical engineering, vol 25. Springer, Berlin Heidelberg New York
2. Axen R, Porath J, Ernback S (1967) Nature 214: 1302
3. Jakoby WB, Wilchek M (eds) (1974) In: Methods in enzymology, vol 34. Academic, New York
4. Lowe CR, Dean PDG (1974) Affinity chromatography. Wiley, New York
5. Lowe CR (1984) In: Wiseman A (ed) Topics in enzyme and fermentation biotechnology, vol 9. Ellis Horwood, Chichester, p 78
6. Scopes RK (1987) Analyt Biochem 165: 235
7. Scawen MD, Atkinson T (1987) In: Clonis YD, Atkinson A, Bruton CJ, Lowe CR (eds) Reactive dyes in protein and enzyme technology. Macmillan, Basingstoke, p 51
8. Lowe CR and Dean PDG (1971) FEBS Lett 14: 313
9. Mosbach K, Guilford H, Olsson R, Scott M (1972) Biochem J 127: 625
10. Haff LA, Easterday RL (1978) In: Sundaram PV, Eckstein F (eds) Theory and practice in affinity techniques. Academic, New York, p 23
11. Lowe CR (1979) In: Work TS, Work E (eds) An Introduction to affinity chromatography. North Holland, Amsterdam
12. Yang C-M, Tsao GT (1984) In: Fiechter A (ed) Advances in biochemical engineering, vol 25. Springer, Berlin Heidelberg New York, p 19
13. Dean PDG, Watson DH (1979) J Chromatogr 165: 301
14. More JE, Hitchcock AG, Price S, Rott J, Harvey MJ (1989) In: Vijayalakshmi MA, Bertrand O (eds) Dye-protein interactions: developments and applications. Elsevier, London, p 265
15. Kopperschläger G, Freyer R, Diezel W, Hofmann E (1968) FEBS Let. 1: 137
16. Haeckel R, Hess B, Lauterborn W, Wüster K-H (1968) Hoppe-Seyler's Z Physiol Chem 349: 699
17. Böhme H-J, Kopperschläger G, Schulz J, Hofmann E (1972) J Chromatogr 69: 209
18. Glazer AN (1970) Proc Natl Acad Sci USA 65: 1057
19. Watson DH, Harvey MJ, Dean PDG (1978) Biochem J 173: 591
20. Biellmann J-F, Samama J-P, Bränden CI, Eklund H (1979) Eur J Biochem 102: 107
21. Thompson ST, Cass KH, Stellwagen E (1975) Proc Natl Acad Sci USA 72: 669

22. Stellwagen E (1977) Acc Chem Res 10: 92
23. Beissner RS, Quiocho FA, Rudolph FB (1979) J Mol Biol 134: 847
24. Angal S, Dean PDG (1978) FEBS Lett 96: 346
25. Subramanian S (1984) CRC Crit Rev Biochem 16: 169
26. Comer MJ, Bruton CJ, Atkinson T (1979) J Appl Biochem 1: 259
27. Quadri F, Dean PDG (1980) Biochem J 191: 53
28. Clonis YD, Goldfinch MJ, Lowe CR (1981) Biochem J 197: 203
29. Clonis YD, Lowe CR (1981) Biochim Biophys Acta 659: 86
30. Atkinson T, Hammond PM, Hartwell DL, Hughes P, Scawen MD, Sherwood RF, Small DAP, Bruton CJ, Harvey MJ, Lowe CR (1981) Biochem Soc Trans 9: 290
31. Hey Y, Dean PDG (1983) Biochem J 209: 363
32. Scopes RK (1984) Analyt Biochem 136: 525
33. Scopes RK (1986) J Chromatogr 376: 131
34. Scopes RK (1987) Protein purification principles and practice, 2nd Ed. Springer, New York
35. Siegel E, Schundehutte K-H, Hildebrand D (1972) In: Venkataraman K (ed) The chemistry of synthetic dyes, vol 6. Academic, New York
36. Stead, CV (1987) In: Clonis YD, Atkinson A, Bruton CJ, Lowe CR (eds) Reactive dyes in protein and enzyme technology. Macmillan, Basingstoke, p 13
37. Stead CV (1989) In: Vijayalakshmi MA, Bertrand O (eds) Dye-protein interactions: developments and applications. Elsevier, London, p 21
38. Burton SJ, McLoughlin SB, Stead CV, Lowe CR (1988) J Chromatogr 435: 127
39. Lowe CR, Pearson JC (1984) In: Jakoby WB (ed) Methods in enzymology, vol 104. Academic, London, p 97
40. Weber B, Willeford K, Moe J, Piszkiewicz D (1979) Biochem Biophys Res Commun 86: 252
41. Hildebrand D (1972) In: Venkataraman K (ed) The chemistry of synthetic dyes, vol 6. Academic, New York
42. Heyns W, DeMoor P (1974) Biochem Biophys Acta 358: 1
43. Baird JK, Sherwood RF, Carr RJG, Atkinson A (1976) FEBS Lett 70: 61
44. Boyer PM, Hsu JT (1992) Chem Eng Sci 47: 241
45. Clonis YD (1987) In: Clonis YD, Atkinson A, Bruton CJ, Lowe CR (eds) Reactive dyes in protein and enzyme technology. Macmillan, Basingstoke, p 33
46. Clonis YD, Stead CV, Lowe CR (1987) Biotechnol Bioeng 30: 621
47. Clonis YD, Lowe CR (1980) Biochem J 191: 247
48. Clonis YD (1988) CRC Crit Rev Biotechnol 7: 263
49. Lowe CR, Burton SJ, Pearson JC, Clonis YD (1986) J Chromatogr 376: 121
50. Farmer EE, Easterby JS (1982) Analyt Biochem 123: 373
51. Goward CR, Scawen MD, Atkinson T (1987) Biochem J 246: 83
52. Scopes RK (1989) In: Vijayalakshmi MA, Bertrand O (eds) Dye-protein interactions: developments and applications. Elsevier, London, p 97
53. Cadelis F, Vijayalakshmi MA, Narayan SR (1989) In: Vijayalakshmi MA, Bertrand O (eds) Dye-protein interactions: developments and applications. Elsevier, London, p 33
54. Burton SJ, Stead CV, Lowe CR (1988) J Chromatogr 455: 201
55. Burton SJ, Stead CV, Lowe CR (1990) J Chromatogr 506: 109
56. Boyer PM, Hsu JT (1989) Biotechnol Tech 4: 61
57. Subramanian S (1989) In: Vijayalakshmi MA, Bertrand O (eds) Dye-protein interactions: developments and applications. Elsevier, London, p 56
58. Bowie JU, Luthy R, Eisenberg D (1991) Science 253: 164
59. Hughes P, Lowe CR, Sherwood RF (1982) Biochim Biophys Acta 700: 90
60. Hughes P, Sherwood RF, Lowe CR (1982) Biochem J 205: 453
61. Hughes P, Sherwood RF, Lowe CR (1984) Eur J Biochem 144: 135
62. Hughes P (1989) In: Vijayalakshmi MA, Bertrand O (eds) Dye-protein interactions: developments and applications. Elsevier, London, p 207
63. Sherwood RF, Melton RG, Alwan SM, Hughes P (1985) Eur J Biochem 148: 447
64. Hughes P, Sherwood RF (1987) In: Clonis YD, Atkinson A, Bruton CJ, Lowe CR (eds) Reactive dyes in protein and enzyme technology. Macmillan, Basingstoke, p 125
65. Hodgson J (1990) Bio/Technol 8: 864
66. Easterday RL, Easterday IM (1974) In: Dunlap RB (ed) Immobilized biochemicals and affinity chromatography. Plenum, New York, p 123 (Advances in medical biology, vol 42)
67. Kroviarski Y, Cochet S, Vadon C, Truskolaski A, Boivin P, Bertrand O (1988) J Chromatogr 449: 403

68. Hondmann DHA, Visser J (1990) J Chromatogr 510: 155
69. Hulak I, Nguyen C, Girot P, Boscetti E (1991) J Chromatogr 539: 355
70. Birkenmeier G, Kopperschläger (1982) J Chromatogr 235: 237
71. Algiman E, Kroviarski Y, Cochet S, Kong YL, Muller D, Dhermy D, Bertrand O (1990) J Chromatogr 510: 165
72. Mislovičová D, Gemeiner P, Kuniak L, Zemek J (1980) J Chromatogr 194: 95
73. Mislovičová D, Gemeiner P, Breier A (1988) Enz Microb Technol 10: 568
74. Mislovičová D, Gemeiner P, Stratilová E, Horváthová M (1990) J Chromatogr 510: 197
75. Reuter R, Naumann M, Kopperschläger (1990) J Chromatogr 510: 189
76. Ruaan R-C, Blair JB, Shaeiwitz JA (1988) Biotechnol Prog 4: 107
77. Anderson PA, Jervis L (1978) Biochem Soc Trans 6: 263
78. Konečný P, Smrž M, Borák J Slováková J (1987) J Chromatogr 398: 387
79. Sturgeon CM, Kennedy JF (1986) In: Zaborsky OR, Kennedy JF (eds) Enzyme and microbial technology, literature survey. Grosvenor, Portsmouth
80. Inman JK (1982) In: Gribnau TCJ, Visser J, Nivard RJF (eds) Affinity chromatography and related techniques. Elsevier, Amsterdam, p 51 (Analytical chemistry symposia series, vol 9)
81. Apps DK, Gleed CD (1976) Biochem J 159: 441
82. Travis J, Bowen J, Tewksbury D, Johnson D, Pannell R (1976) Biochem J 157: 301
83. Wilson JE (1976) Biochem Biophys Res Commun 72: 816
84. Clonis YD (1987) Bio/Technol 5: 1290
85. Hammond PM, Atkinson T, Scawen MD (1986) J Chromatogr 366: 79
86. Liu YC, Stellwagen E (1986) J Chromatogr 376: 149
87. Liu YC, Stellwagen E (1987) J Biol Chem 262: 583
88. Hogg PJ, Winzor DJ (1985) Arch Biochem Biophys 240: 70
89. Mayes AG, Moore JD, Eisenthal R, Hubble J (1990) Biotechnol Bioeng 36: 1090
90. Vijayalakshmi MA (1989) In: Vijayalakshmi MA, Bertrand O (eds) Dye-protein interactions: developments and applications. Elsevier, London, p 337
91. Lowe CR, Burton N, Dilmaghanian S, McLoughlin S, Pearson J, Stewart D, Clonis YD (1989) In: Vijayalakshmi MA, Bertrand O (eds) Dye-protein interactions: developments and applications. Elsevier, London, p 11
92. Pearson JC, Lowe CR (1985) in Abst 6th Int Symp Affinity Chromatogr Related Techniques, Prague, p 47
93. Jones K (1988) Chromatographia 25: 443
94. Stewart DJ, Purvis DR, Lowe CR (1990) J Chromatogr 510: 177
95. Danielson ND, Beaver LG, Wangsa J (1991) J Chromatogr 544: 187
96. Roy SK, Nishikawa AH (1979) Biotechnol Bioeng 21: 775
97. Scopes RK, Griffiths-Smith K (1984) Analyt Biochem 136: 530
98. Subramanian S (1987) Prep Biochem 17: 297
99. Jungblot P, Klose J (1989) J Chromatogr 482: 125
100. Unitika Ltd (1985) Jpn Kokai Tokkyo Koho, JP 60 78578, p 367
101. Kenny C, Moschera JA, Stein S (1981) In: Pestka S (ed) Methods in enzymology, vol 78. Academic Press, New York, p 437
102. Byfield PGH (1989) In: Vijayalakshmi MA, Bertrand O (eds) Dye-protein interactions: developments and applications. Elsevier, London, p 244
103. Brehm RD, Tranter HS, Hambleton P, Melling J (1990) Appl Env Microb 56: 1067
104. Jankowski W, Van Muenchkausen W, Sulkowski E, Carter E (1976) Biochemistry 15: 5182
105. Kroviarski Y, Cochet S, Vadon C, Truskolaski A, Boivin P, Bertrand O (1988) J Chromatogr 449: 413
106. Scopes RK, Testolin V, Stoter A, Griffiths-Smith K, Algar EM (1985) Biochem J 228: 627
107. Goward CR, Hartwell R, Atkinson T, Scawen MD (1986) Biochem J 237: 415
108. Scawen MD, Hammond PM, Comer MJ, Atkison T (1983) Analyt Biochem 132: 413
109. Tsai AM, Englert D, Graham EE (1990) J Chromatogr 504: 89
110. Arve BH, Liapis AI (1988) Biotechnol Bioeng 32: 616
111. Cowen GH, Gosling IS, Laws JF, Sweetenham WP (1986) J Chromatogr 363: 37
112. Chase HA (1984) J Chromatogr 297: 179
113. Horstmann BJ, Kenney CN, Chase HA (1986) J Chromatogr 361: 179
114. Horstmann BJ, Chase HA (1989) Chem Eng Res Des 67: 243
115. Skidmore GL, Horstmann BJ, Chase HA (1990) J Chromatogr 498: 113
116. Arnold FH, Blanch HW, Wilke CR (1985) Chem Eng J 30: B9
117. Arnold FH, Blanch HW, Wilke CR (1985) Chem Eng J 30: B25

118. Arnold FH, Blanch HW (1986) J Chromatogr 355: 13
119. Anspach FB, Johnston A, Wirth H-J, Unger KK, Hearn MTW (1989) J Chromatogr 476: 205
120. Anspach FB, Johnston A, Wirth H-J, Unger KK, Hearn MTW (1990) J Chromatogr 499: 103
121. Graves DJ, Wu YT (1974) In: Jakoby WB, Wilchek M (eds) Methods in enzymology, vol 34. Academic, New York, p 140
122. Walsh C (1979) Enzymatic reaction mechanisms. Freeman, San Francisco, p 123
123. Pompon D, Guiard B, Lederer F (1980) Eur J Biochem 110: 565
124. Dunn BM, Chaiken IM (1975) Biochemistry 14: 2343
125. Dunn BM, Chaiken IM (1974) Proc Natl Acad Sci USA 71: 2382
126. Nichol LW, Ogston AG, Winzor DJ, Sawyer WH (1974) Biochem J 143: 435
127. Hethcote HW, DeLisi C (1982) J Chromatogr 248: 183
128. Hogg PJ, Winzor DG (1984) Arch Biochem Biophys 234: 55
129. Liu YC, Ledger R, Stellewagen E (1984) J Biol Chem 259: 3796
130. Herak DC, Merrill EW (1989) Biotechnol Prog 5: 9
131. Herak DC, Merrill EW (1990) Biotechnol Prog 6: 33
132. Krause S, Kroner KH, Deckwer W-D (1991) Biotechnol Tech 5: 199
133. Tsou H-S, Graham EE (1985) AIChE J 31: 1959
134. Sada E, Katoh S, Sukai K, Tohma M, Kondo A (1986) Biotechnol Bioeng 28: 1497
135. Liapis AI (1989) J Biotechnol 11: 143
136. Thomas HC (1944) J Amer Che Soc 66: 1664
137. Johnston A, Hearn MTW (1990) J Chromatogr 512: 101
138. Hiester NK, Vermeulen T (1952) Chem Eng Prog 48: 505
139. Skidmore GL, Chase HA (1990) J Chromatogr 505: 329
140. Tomasselli AG, Schirmes RH, Noda LH (1979) Eur J Biochem 93: 257
141. Tomasselli AG, Noda LH (1980) Eur J Biochem 103: 481
142. Scawen MD, Darbyshire J, Harvey MJ, Atkinson T (1982) Biochem J 203: 669
143. Kulbe KD, Schuer R (1979) Analyt Biochem 93: 46
144. Ito Y, Tamasselli AG, Noda LH (1980) Eur J Biochem 105: 85
145. Baydoun HJ, Hoppe W, Friest W, Wagner KG (1982) J Biol Chem 257: 1032
146. Knight P (1990) Bio/Technol 8: 200
147. Neale AD, Scopes RK, Kelley JM, Wettenhall REH (1986) Eur J Biochem 154: 119
148. Clonis YD, Small DAP (1987) In: Clonis YD, Atkinson A, Bruton CJ, Lowe CR (eds) Reactive dyes in protein and enzyme technology. Macmillan, Basingstoke, p 87
149. Ohlson S, Hansson L, Larsson P-O, Mosbach K (1978) FEBS Lett 93: 5
150. Glad M, Ohlson S, Hansson L, Månsson M-O, Mosbach K (1980) J Chromatogr 200: 254
151. Sportsman JR, Wilson GS (1980) Analyt Chem 52: 2013
152. Lowe CR, Glad M, Larsson P-O, Ohlson S, Small DAP, Atkinson T, Mosbach K (1981) J Chromatogr 215: 303
153. Small DAP, Atkinson T, Lowe CR (1981) J Chromatogr 216: 175
154. Small DAP, Atkinson T, Lowe CR (1983) J Chromatogr 266: 151
155. Clonis YD, Jones K, Lowe CR (1986) J Chromatogr 363: 31
156. Kadushevichyus VA, Sudzhyuvene OF, Peslyakas II (1986) Appl Biochem Microbiol 22: 237
157. Baskheviciute BB, Sudzhyuvene OF, Peslyakas II, Glemza AA, Migunov VN, Lyakhova TD, Pozina IM (1987) Appl Biochem Microbiol 23: 493
158. Clonis YD (1987) J Chromatogr 407: 179
159. Anspach B, Unger KK, Davies J, Hearn MTW (1988) J Chromatogr 457: 195
160. Livingston AG, Chase HA (1989) J Chromatogr 481: 159
161. Clonis YD, Lowe CR (1991) J Chromatogr 540: 103
162. Mislovičová D, Novák I, Pašteka M (1991) J Chromatogr 543: 9
163. Regnier FE, Noel R (1976) J Chromatogr Sci 14: 316
164. Johansson G (1987) In: Clonis YD, Atkinson A, Bruton CJ, Lowe CR (eds) Reactive dyes in protein and enzyme technology. Macmillan, Basingstoke, p 101
165. Kopperschläger G, Lorenz G, Usbeck E (1983) J Chromatogr 259: 97
166. Johansson G, Joelsson M (1985) Biotechnol Bioeng 27: 621
167. Kopperschläger G, Johansson G (1982) Analyt Biochem 124: 117
168. Johansson G, Joelsson M (1985) Enzyme Microb Technol 7: 629
169. Cordes A, Kula M-R (1986) J Chromatogr 376: 375
170. Tjerneld F, Johansson G, Joelsson M (1987) Biotechnol Bioeng 30: 809
171. Kula M-R, Kroner KH, Hustedt H (1982) In: Fiechter A (ed) Advances in biochemical engineering, vol 24. Springer, Berlin Heidelberg New York, p 73

172. Michaels AS, Matson SL (1985) Desalination 53: 231
173. Scawen MD (1985) Biochem J 203: 699
174. Unarska M, Davies PA, Esnouf MP, Bellhouse BJ (1990) J Chromatogr 519: 53
175. Champluvier B, Kula M-R (1991) J Chromatogr 539: 315
176. Schisla DK, Carr PW, Cussler EL (1990) Ext Abst AIChE 1990 Ann Mtg, Paper 68A
177. Brandt S, Goffe RA, Kessler SB, O'Connor JL, Zale SE (1988) Bio/Technol 6: 779
178. Luong JHT, Nguyen A-L, Male KB (1987) Bio/Technol 5: 564
179. Mattiasson B, Ramstorp M (1983) Ann NY Acad Sci 13: 307
180. Ling TGI, Mattiasson B (1989) Biotechnol Bioeng 34: 1321
181. Luong JHT, Male KB, Nguyen A-L (1988) Biotechnol Bioeng 31: 439
182. Bertrand O, Cochet S, Kroviarski Y, Truskolaski A, Boivin P (1985) J Chromatogr 346: 111
183. Pearson JC, Burton SJ, Lowe CR (1986) Analyt Biochem 158: 382
184. Riahi B, Vijayalakshmi (1989) In: Vijayalakshmi MA, Bertrand O (eds) Dye-protein inter-
     actions: developments and applications. Elsevier, London, p 197
185. Pearson JC (1987) In: Clonis YD, Atkinson A, Bruton CJ, Lowe CR (eds) Reactive dyes in
     protein and enzyme technology. Macmillan, Basingstoke, p 187
186. Clonis YD (1987) In: Clonis YD, Atkinson A, Bruton CJ, Lowe CR (eds) Reactive dyes in
     protein and enzyme technology. Macmillan, Basingstoke, p 193

Modeling of Nonlinear Multicomponent Chromatography

T. Gu[1], G.-J. Tsai[2], and G. T. Tsao[3]
[1] Department of Chemical Engineering, Ohio University, Athens, OH 45701, USA
[2] Building 130, Lederle Laboratories, Pearl River, NY 10965, USA
[3] Laboratory of Renewable Resources Engineering, 1295 Potter Center, Purdue University, West Lafayette, IN 47907-1295, USA

Advances in Biochemical Engineering
Biotechnology, Vol. 49
Managing Editor: A. Fiechter
© Springer-Verlag Berlin Heidelberg 1993

In the age of rapid development of biotechnology, preparative and large scale chromatography becomes more and more popular. Unlike analytical chromatography, dispersion and mass transfer effects are often significant in preparative and large scale chromatography. Concentration overload often leads to nonlinearity of the system. The study of nonlinear chromatography becomes more and more demanding. Much work has been done in the past two decades, but many topics of practical importance still need to be tackled. In this chapter, a review is given on different models for chromatography. This chapter provides a brief review of different mathematical models for nonlinear chromatography. A general multicomponent rate model, which accounts for various mass transfer mechanisms and nonlinear isotherms is presented. This comprehensive model is a very powerful tool for the study of the dynamics of nonlinear multicomponent chromatography. This chapter also presents an efficient numerical method for the solution of the model and its numerous extensions. As an example, the model is used for the study of some interesting effects of isotherm characteristics of the displacer on the optimization of stepwise displacement.

List of Symbols

Symbol	Description
a_i	constant in Langmuir isotherm for component i, $b_i C_i^\infty$
b_i	adsorption equilibrium constant for component i, k_{ai}/k_{di}
Bi_i	Biot number of mass transfer for component i, $k_i R_p/(\varepsilon_p D_{pi})$ or $k_i R_p/(\varepsilon_{pi}^a D_{pi})$
C_{bi}	bulk phase concentration of component i
C_{fi}	feed concentration profile of component i, a time dependent variable
C_{0i}	concentration used for nondimensionalization, $\max\{C_{fi}(t)\}$
C_{pi}	concentration of component i in the stagnant fluid phase inside particle macropores
C_{pi}^s	concentration of component i in the solid phase of particle (mole adsorbate/unit volume of particle skeleton)
C_i^∞	adsorption saturation capacity for component i (mole adsorbate/unit volume of particle skeleton)
$\bar{C}$	adsorption saturation capacity based on the unit volume of the bed
$\bar{C}_i$	concentration of component i in the stationary phase based on the unit volume of the bed
C_i	concentration of component i in the fluid phase based on the unit volume of the bed
c_{bi}	$= C_{bi}/C_{0i}$
c_{pi}	$= C_{pi}/C_{0i}$
c_{pi}^s	$= C_{pi}^s/C_{0i}$
c_i^∞	$= C_i^\infty/C_{0i}$
D_{bi}	axial or radial dispersion coefficient of component i
Da_i^a	Damköhler number for adsorption, $\dfrac{L(k_{ai}C_{0i})}{v}$
Da_i^d	Damköhler number for desorption, Lk_{di}/v
F_i^{ex}	size exclusion factor for component i, $\varepsilon_{pi}^a/\varepsilon_p$

k_i	film mass transfer coefficient of component i
k_{ai}	adsorption rate constant for component i
k_{di}	desorption rate constant for component i
L	column length
N	number of interior collocation points
Ne	number of quadratic elements
Ns	number of components
Pe_{Li}	Peclet number of axial dispersion for component i, vL/D_{bi}
R	radial coordinate for particle
R_p	particle radius
r	$= R/R_p$
t	time
v	interstitial velocity
Z	axial coordinate
z	$= Z/L$

Greek Letters

ε_b	bed void volume fraction
ε_p	particle porosity
ε_{pi}^a	accessible particle porosity of component i
η_i	dimensionless constant, $\dfrac{\varepsilon_p D_{pi} L}{R_p^2 v}$
ξ_i	dimensionless constant for component i, $3Bi_i \eta_i (1 - \varepsilon_b)/\varepsilon_b$
τ	dimensionless time, $\dfrac{vt}{L}$
τ_{imp}	dimensionless time duration for a rectangular pulse of the sample
ϕ	Lagrangian interpolation function

1 Review of Models for Chromatography

A very comprehensive review on the dynamics and mathematical modeling of adsorption and chromatography was given by Ruthven [1]. Models in this area are generally classified into three categories [1]: equilibrium theory, plate models and rate models.

1.1 Equilibrium Theory

Glueckauf [2] is considered as being the first person to develop the equilibrium theory of multicomponent isothermal adsorption [1]. The theory further developed into the interference theory by Helfferich and Klein [3] is mainly aimed at stoichiometric ion-exchange systems with constant separation factors. A mathematically parallel treatise for systems with multicomponent Langmuir isotherms was developed by Rhee and co-workers [4, 5].

Equilibrium theory assumes direct local equilibrium between the mobile phase and the stationary phase, neglecting axial dispersion and mass transfer resistances. The theory gives good interpretation of experimental results for chromatographic columns with fast mass transfer rates shown by many analytical and some preparative columns. It can provide general location of the concentration profiles of a chromatographic system but fails to provide accurate details if mass transfer effects in the system are significant [6]. Equilibrium theory has been widely used for the study of multicomponent interference effects [3] and ideal displacement development [5]. Many cases of practical application have been reported [3, 7–12].

1.2 Plate Models

Generally speaking, there are two kinds of plate models, which may also be called staged models or staged theories [13]. The first kind is directly analogous to the "tanks in series" model for nonideal flow systems [1]. In such a model, the column is divided into a series of small artificial elements. Inside each element the content is assumed to be completely mixed. This gives a set of first order ordinary differential equations (ODE's) that describe the adsorption and interfacial mass transfer processes. Many researchers have contributed to this kind of plate model [1, 14–16]. However, plate models of this kind generally are not suitable for multicomponent chromatography since the equilibrium stages may not be assumed equal for different components.

The other kind of plate model is formulated based on the distribution factors which determine the equilibrium of each component in each of the artificial stages, and the model solution involves recursive iterations, rather than solving for ODE systems. The most popular of this kind are the Craig distribution models. Considering the blockage effect, the Craig models are applicable to multicomponent systems. Descriptions of Craig models were given by Eble et al. [17], Seshadri and Deming [18], and Solms et al. [19]. In recent years, Craig models have been extensively used for the study of column overload problems [17, 20].

1.3 Rate Models

The word "rate" refers to the rate expression or rate equation for the mass transfer between the mobile phase and the stationary phase. A rate model usually consists of two sets of differential mass balance equations, one for the bulk-fluid phase, the other for the particle phase. Different rate models have different complexities. A comprehensive review of rate models was given by Ruthven [1].

1.3.1 Rate Expressions and Particle Phase Governing Equations

The solid film resistance hypothesis was first proposed by Glueckauf and Coates [21]. It assumes a linear driving force between the equilibrium concentrations in the stationary phase (determined from the isotherm) and the average fictitious concentrations in the stationary phase. Because of its simplicity, this rate expression has been used by many researchers [1, 22–25] but this model cannot provide the details of the mass transfer processes.

The fluid film resistance mechanism which also assumes a linear driving force is widely used [1]. It is often called external mass transfer resistance. If the concentration gradient inside the particle phase is ignored, the model then becomes the lumped particle model, which has been used by some researchers [27–29]. If the Biot number for mass transfer, which reflects the ratio of the characteristic rate of film mass transfer over that of intraparticle diffusion, is much larger than 1, the external film mass transfer resistance can be neglected with respect to pore diffusion.

In many cases both external mass transfer and intraparticle diffusion must be considered. A local equilibrium is often assumed between the concentration in the stagnant fluid phase inside macropores and the solid phase of the particle. Such a rate mechanism is adequate to describe the adsorption and mass transfer between the bulk-fluid phase and the particle phase, and inside the particle phase in most chromatographic processes. The local equilibrium assumption here is different from that made for the equilibrium model which assumes a direct equilibrium of concentrations in the solid and the liquid phase without any kind of mass transfer resistances.

1.3.2 Adsorption Kinetics and Affinity Chromatography

In some cases, the adsorption and desorption rates may not be high enough and the assumption of the local equilibrium between the concentration in the stagnant fluid phase inside macropores and the solid phase of the particle is no longer valid. Kinetic models must be used. Some kinetic models were reviewed by Ruthven [1] and Lee et al. [30]. The second order kinetics has been widely used in affinity chromatography [31–39]. If the saturation capacities for all the solutes are the same, the second order kinetics reduces to the Langmuir isotherm when equilibrium is assumed.

1.3.3 Governing Equations for Bulk-Fluid Phase

The partial differential equation for the bulk-fluid phase can be easily obtained with differential mass balances. They usually contain the following terms: axial dispersion, convection, transient, and the interfacial flux. Such equations themselves are generally linear if physical parameters are not concentration dependent. They become nonlinear when coupled with nonlinear rate expressions.

Analytical solutions may be obtained using Laplace transformation [39, 40] for many isothermal, single component systems with linear isotherms. The linear operator method [41] can also be used to solve problems in linear chromatography. For more complex systems, especially those involving nonlinear isotherms, analytical solutions generally cannot be derived [1]. With the rapid growth of the availability of fast and powerful computers and development of efficient numerical methods, it is now possible to obtain numerical solutions to complex rate models that consider various forms of mass transfer mechanisms [42]. Complex rate models are now becoming more and more popular especially in the study of preparative and large scale chromatography.

1.4 General Multicomponent Rate Models

A rate model which considers axial dispersion, external mass transfer, intraparticle diffusion and nonlinear isotherms is considered a general multicomponent rate model. Such a general model is adequate in most cases to describe the adsorption and mass transfer processes in multicomponent chromatography. In some cases surface adsorption and size exclusion, adsorption kinetics, etc., may have to be included to give an adequate account for a particular system.

Several groups of researchers have proposed and solved various general multicomponent rate models using different numerical approaches [42–45].

1.5 Solution to the General Multicomponent Rate Models

A general multicomponent rate model consists of a coupled PDE system with two sets of mass balance equations in the bulk-fluid and particle phases for each component, respectively. The transient PDE system becomes nonlinear if any nonlinear isotherms or nonlinear kinetics are involved in the system.

The finite difference method is a very simple numerical procedure and can be directly applied for the solution to the model [45, 46], but this procedure often requires a huge amount of memory space, and its efficiency and accuracy are not competitive compared with other advanced numerical methods, such as orthogonal collocation (OC), finite element, and orthogonal collocation on finite element (OCFE).

The general strategy for solving a nonlinear transient PDE system numerically using the advanced numerical methods is to discretize the spatial axes in the model equations first, and then solve the resulting ODE system using an ODE solver.

1.5.1 Discretization of Particle Phase Equations

The OC method is a very accurate, efficient and simple method for discretization. It has been widely used for particle problems [47, 48] and is obviously the best choice for the particle phase governing equations of general multicomponent rate models [42, 43, 44].

1.5.2 Discretization of Bulk-Fluid Phase Equations

Concentration gradients in the bulk-fluid phase can be very stiff, thus, the OC method is no longer suitable, since global splines using high order polynomials are very expensive [48] and sometimes not stable. The OCFE method uses linear elements for global spline and collocation points inside each element. No numerical integration for element matrices is needed because of the use of linear elements. This discretization method can be used for systems with stiff gradients [48].

The finite element method with a higher order of interpolation functions (typically quadratic, or occasionally cubic) is a very powerful method for stiff systems. Its highly streamlined structure provides unsurpassed convenience and versatility. This method is especially useful for systems with variable physical parameters, such as radial flow chromatography and nonisothermal adsorption with or without chemical reaction. Chromatography of some biopolymers also involves variable axial dispersion coefficient [49].

1.5.3 Solution to the ODE System

If the finite element method is used for the discretization of bulk-fluid phase equations and OC for the particle phase equations, an ODE system then results. The ODE system with initial values can be readily solved using an ODE solver such as subroutine IVPAG of the IMSL [50], which uses the powerful Gear's stiff method [51].

If the discretization of the bulk-fluid phase equations is carried out using the OCFE method, an ODE system coupled with some algebraic equations which come from the continuity of boundary fluxes results [44, 48]. The system can be solved using an available differential algebraic equation solver.

Such a system can also be conveniently solved with an ODE solver if one manipulates the user-supplied function subroutine which evaluates the concentration derivatives for the ODE solver to eliminate those algebraic equations in an in situ fashion. This is possible since the trial concentrations are given as arguments for the subroutine. This approach helps reduce the total number of equations in the final system. It was apparently adopted by Gardini et al. [52], for a multi-phase reaction engineering problem.

2 A General Multicomponent Rate Model for Axial Flow Chromatography

2.1 Model Assumptions

Figure 1 shows the anatomy of a chromatographic column with axial flow. The following basic assumptions are needed for the formulation of the general rate model.

- The multicomponent fixed-bed process is isothermal.
- The bed is packed with porous adsorbents which are spherical and uniform in size.
- The concentration gradients in the radial direction of the bed are negligible.
- Local equilibrium exists for each component between the pore surface and the stagnant fluid phase in the macropores.
- The diffusional and mass transfer coefficients are constant and independent of the mixing effects of the components.

2.2 Model Formulation

Based on these basic assumptions, the following governing equations can be formulated from the differential mass balances for each component in the bulk-fluid and the particle phases.

$$- D_{bi} \frac{\partial^2 C_{bi}}{\partial Z^2} + v \frac{\partial C_{bi}}{\partial Z} + \frac{\partial C_{bi}}{\partial t} + \frac{3 k_i (1 - \varepsilon_b)}{\varepsilon_b R_p} (C_{bi} - C_{pi, R = R_p}) = 0 \qquad (1)$$

$$(1 - \varepsilon_p) \frac{\partial C_{pi}^s}{\partial t} + \varepsilon_p \frac{\partial C_{pi}}{\partial t} - \varepsilon_p D_{pi} \left[\frac{1}{R^2} \frac{\partial}{\partial R} \left(R^2 \frac{\partial C_{pi}}{\partial R} \right) \right] = 0 \qquad (2)$$

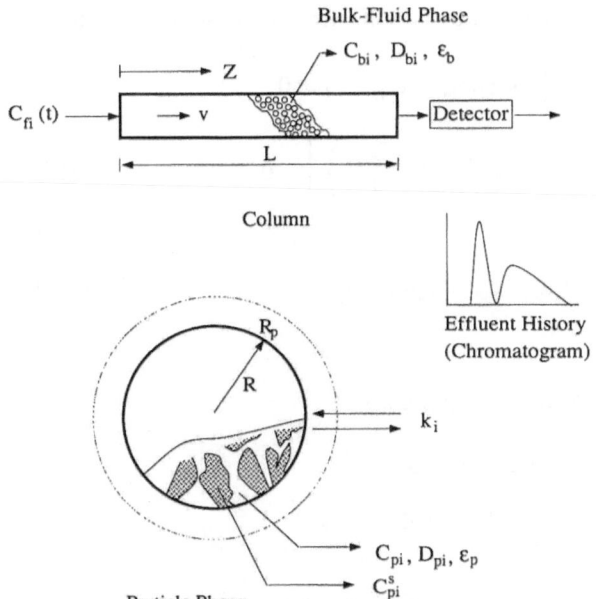

Fig. 1. Anatomy of a chromatographic column

with the initial and boundary conditions

$$t = 0, \quad C_{bi} = C_{bi}(0,Z), \; C_{pi} = C_{pi}(0,R,Z) \tag{3, 4}$$

$$Z = 0, \quad \frac{\partial C_{bi}}{\partial Z} = \frac{v}{D_{bi}}(C_{bi} - C_{fi}(t)) \quad Z = L, \quad \frac{\partial C_{bi}}{\partial Z} = 0 \tag{5, 6}$$

$$R = 0, \quad \frac{\partial C_{pi}}{\partial R} = 0 \qquad R = R_p, \quad \frac{\partial C_{pi}}{\partial R} = \frac{k_i}{\varepsilon_p D_{pi}}(C_{bi} - C_{pi, R = R_p}) \tag{7, 8}$$

Equations (1) and (2) are coupled via $C_{pi}, \; R = R_p$ which is the concentration of component i at the surface of a particle. In Eq. (2), C_{pi}^s is the concentration of component i in the solid phase of the adsorbents based on the unit volume of the solid, excluding pores. It is directly linked to the multicomponent isotherms which couple the PDE system based on assumption (4). Concentrations C_{bi} and C_{pi} are based on the unit volume of mobile phase fluid. The effective diffusivities, D_{pi}, in this work do not include the particle porosity.

By introducing the following dimensionless terms

$$c_{bi} = C_{bi}/C_{0i}, \; c_{pi} = C_{pi}/C_{0i}, \; c_{pi}^s = C_{pi}^s/C_{0i}, \; r = R/R_p, \; z = Z/L, \; \tau = vt/L$$

$$Pe_{LI} = vL/D_{bi}, \; Bi_i = k_i R_p/\varepsilon_p D_{pi}, \; \eta_i = \varepsilon_p D_{pi} L/R_p^2 v, \; \xi_i$$

$$= 3Bi_i \eta_i(1 - \varepsilon_b)/\varepsilon_b$$

the PDE system can be transformed into the following dimensionless forms.

$$-\frac{1}{Pe_{Li}}\frac{\partial^2 c_{bi}}{\partial z^2} + \frac{\partial c_{bi}}{\partial z} + \frac{\partial c_{bi}}{\partial \tau} + \xi_i(c_{bi} - c_{pi,\, r=1}) = 0 \tag{9}$$

$$\frac{\partial}{\partial \tau}[(1 - \varepsilon_p)c_{pi}^s + \varepsilon_p c_{pi}] - \eta_i\left[\frac{1}{r^2}\frac{\partial}{\partial r}\left(r^2\frac{\partial c_{pi}}{\partial r}\right)\right] = 0 \tag{10}$$

I. C.

$$\tau = 0, \qquad c_{bi} = c_{bi}(0, z), \quad c_{pi} = c_{pi}(0, r, z) \tag{11, 12}$$

B. C.

$$z = 0, \qquad \frac{\partial c_{bi}}{\partial z} = Pe_{Li}(c_{bi} - C_{fi}(\tau)/C_{0i}) \tag{13}$$

For frontal adsorption $C_{fi}(\tau)/C_{0i} = 1$

For elution $C_{fi}(\tau)/C_{0i} = \begin{cases} 1 & 0 \le \tau \le \tau_{imp} \\ 0 & \text{else} \end{cases}$

After the sample introduction (in the form of frontal adsorption):

if component i is displaced, $C_{fi}(\tau)/C_{0i} = 0$

if component i is a displacer, $C_{fi}(\tau)/C_{0i} = 1$

$$z = 1, \qquad \frac{\partial c_{bi}}{\partial z} = 0 \tag{14}$$

$$r = 0, \qquad \frac{\partial c_{pi}}{\partial r} = 0 \quad r = 1, \quad \frac{\partial c_{pi}}{\partial r} = Bi_i(c_{bi} - c_{pi,\, r=1}) \tag{15, 16}$$

Note that all the dimensionless concentrations are based on C_{0i} which is equal to the maximum of the feed profile $C_{fi}(\tau)$. For example, in an elution, if component i is a sample solute in the sample which is injected as a rectangular pulse, the profile of $C_{fi}(\tau)$ is then of a rectangular shape, and its upper boundary value is the value of C_{0i}.

3 Numerical Solution to the Model

An efficient and robust numerical procedure has been developed by Gu et al. [42] for the solution to the above PDE system. It involves two parts. First, the spatial axes, z and r, are discretized. And then the resulting ODE system (with initial values) is solved with an ODE solver (integrator). An overview of the general strategy for the solution is shown in Figure 2.

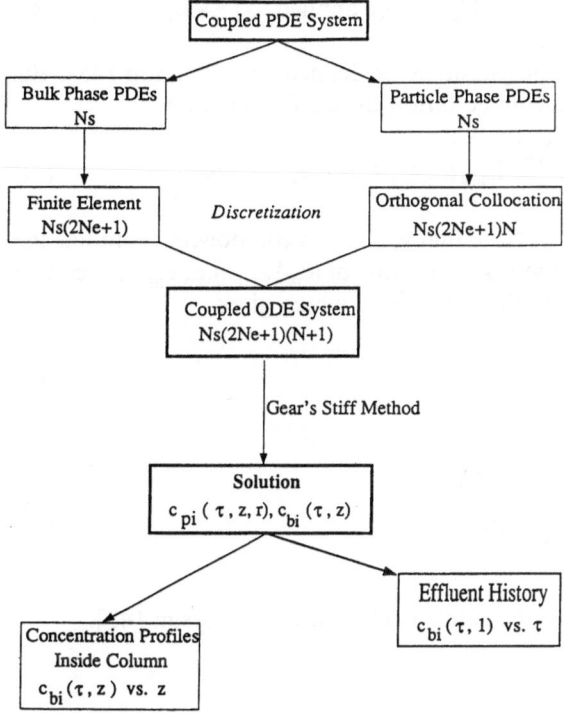

Fig. 2. Solution strategy for the general multicomponent rate model

3.1 Discretization

Equations (9) and (10) can be discretized into a set of ODE's by the finite element and the OC methods respectively. Using the Galerkin approximation and the first weak form [53], Eq. (9) becomes

$$[DB_i][c'_{bi}] + [AKB_i][c_{bi}] = [PB_i] + [AFB_i] \tag{17}$$

where

$$(DB_i)^e_{m,n} = \int \phi_m \phi_n \, dz \tag{18}$$

$$(AKB_i)^e_{m,n} = \int \left(\frac{1}{Pe_{Li}} \frac{\partial \phi_m}{\partial z} \frac{\partial \phi_n}{\partial z} + \phi_m \frac{\partial \phi_n}{\partial z} + \xi_i \phi_m \phi_n \right) dz \tag{19}$$

$$(AFB_i)^e_m = \int \xi_i \phi_m c_{pi, r=1} \, dz \tag{20}$$

in which m, $n \in \{1, 2, 3\}$, and the superscript e indicates that the finite element matrices and vectors are evaluated over each individual element before global assembly. Four point Gauss-Legendre quadratures [53] are used for integrations. The superscript prime in this work indicates a first order time derivative. The bold face variables indicate matrices or vectors. The natural boundary condition $(\mathbf{PB_i})|_{z=0} = -c_{bi} + C_{fi}(\tau)/C_{0i}$ will be applied to $[\mathbf{AKB_i}]$ and $[\mathbf{AFB_i}]$ at $z = 0$. $(\mathbf{PB_i}) = 0$ anywhere else.

Using the same symmetric polynomials as defined by Finlayson [48], Eq. (10) is transformed to the following equation by the OC method.

$$\left(\sum_{j=1}^{N_s} \frac{\partial g_i}{\partial c_{pj}} \frac{dc_{pj}}{d\tau} \right)_1 = \eta_i \sum_{k=1}^{N+1} \mathbf{B}_{1,k}(c_{pi})_k, \quad 1 = 1, 2, \ldots, N \tag{21}$$

in which $g_i = (1 - \varepsilon_p)c_{pi}^s + \varepsilon_p c_{pi}$. Note that g_i for each component i contains the nonlinear multicomponent isotherms. The value of $(c_{pi})_{N+1}$ (i.e., $c_{pi,r=1}$) can be obtained from the boundary condition at $r = 1$, which gives

$$\sum_{j=1}^{N+1} \mathbf{A}_{N+1,j}(c_{pi})_j = Bi_i(c_{bi} - c_{pi,r=1}) \tag{22}$$

or

$$c_{pi,r=1} = \frac{Bi_i c_{bi} - \sum_{j=1}^{N} \mathbf{A}_{N+1,j}(c_{pi})_j}{\mathbf{A}_{N+1,N+1} + Bi_i} \tag{23}$$

where the matrices $\mathbf{A}$ and $\mathbf{B}$ are the same as defined by Finlayson [48].

3.2 Solution to the ODE System

If Ne quadratic elements (i.e. $(2Ne + 1)$ nodes) are used for the z-axis in bulk-fluid phase equations and N interior OC points are used for the r-axis in particle phase equations, the above discretization procedure gives $N_s (2Ne + 1)$ $(N + 1)$ ODE's which are then solved simultaneously by Gear's stiff method [50]. A function subroutine must be supplied to the ODE solver to evaluate concentration derivatives at each element node and OC point with given trial concentration values. The concentration derivatives at each element node (c'_{bi}) are easily determined from Eq. (17). The concentration derivatives at each OC point (c'_{pi}) are coupled because of the complexity of the isotherms which are related to g_i via c_{pi}^s. At each interior OC point, Eq. (21) can be rewritten in the following matrix form.

$$[\mathbf{GP}][\mathbf{c'_p}] = [\mathbf{RH}] \tag{24}$$

where

$$\mathbf{GP}_{ij} = \frac{\partial g_i}{\partial c_{pj}}, c'_{pj} = \frac{dc_{pj}}{d\tau}, \mathbf{RH}_i = \text{right hand side of Eq. (21)}$$

Since the matrices $[\mathbf{GP}]$ and $[\mathbf{RH}]$ are known with given trial concentrations at each interior OC point, the vector $[\mathbf{c'_p}]$ can be easily determined from Eq. (24). Using this approach, we can deal with complex nonlinear isotherms here without any iteration.

3.3 Isotherm Expressions

The numerical procedure discussed above can accommodate any type of nonlinear isotherms as long as they do not cause mathematical singularities. The following two common types of isotherms are used in this work.

(1) Langmuir Isotherm

$$C_{pi}^s = \frac{a_i C_{pi}}{1 + \sum\limits_{j=1}^{Ns} b_j C_{pj}} \quad \text{i.e.,} \quad c_{pi}^s = \frac{a_i c_{pi}}{1 + \sum\limits_{j=1}^{Ns} (b_j C_{0j}) c_{pj}} \tag{25}$$

where $a_i = C^\infty b_i$. Note that $b_j C_{0j}$ can be treated as a dimensionless group for each component. With this the entire model system can then be treated with dimensionless parameters alone. This helps reduce the total number of parameters involved in discussions.

(2) Stoichiometric Isotherm with Constant Separation Factors

$$\bar{C}_i = \frac{\bar{C} C_i}{\sum\limits_{j=1}^{Ns} \alpha_{ji} C_j} = \frac{\alpha_{i,Ns} \bar{C} C_i}{\sum\limits_{j=1}^{Ns} \alpha_{j,Ns} C_j} \tag{26a}$$

where $\alpha_{ij} = 1/\alpha_{ji} = \alpha_{ik} \alpha_{kj}$, and $\alpha_{ii} = 1$. C_i is the concentration of ion component i in the stagnant fluid inside particles. $\bar{C}$ is the saturation capacity and is considered equal for all components. $\bar{C}_i$ is the concentration of ion component i in the solid of the particles. This type of isotherm is widely used in ion exchange and all the concentrations are based on the unit volume of the column rather than on the respective phases as in the case of Langmuir isotherms [3].

The stoichiometric isotherm can be converted into the isotherm shown below, which is the same algebraic expression as the Langmuir isotherm except that the "$1 +$" in the denominator of the Langmuir isotherm expression is dropped.

$$C_{pi}^s = \frac{a_i C_{pi}}{\sum\limits_{j=1}^{Ns} b_j C_{pj}} \quad \text{i.e.,} \quad c_{pi}^s = \frac{(a_i/C_{0i}) c_{pi}}{\sum\limits_{j=1}^{Ns} b_j c_{pj} C_{0j}/C_{0i}} \tag{26b}$$

The following relationships are needed for the conversion.

$$b_i = \alpha_{i,Ns} \quad \text{and} \quad a_i = b_i C^\infty = \frac{\alpha_{i,Ns} \bar{C}}{(1 - \varepsilon_b)(1 - \varepsilon_p)} \quad (i = 1, 2, \ldots, Ns)$$

where ion component Ns is assigned as the basis of the separation factors. Note that the units of a_i and b_i in the Langmuir isotherm and the converted stoichiometric isotherm are not the same. In the stoichiometric isotherm, the concentrations of components cannot all be zero at the same time, which means that the column is never "empty."

4 Efficiency and Robustness of the Numerical Procedure

The solution to the rate model provides the effluent history and the moving concentration profiles inside the column for each component. The concentration profile of each component inside the stagnant fluid phase and the solid phase of the particle can also be obtained, but they are rarely used for discussions.

Generally speaking, one interior collocation point (N = 1) is sufficient in some cases, while more often N = 2 is needed, especially when D_{pi} values are small, which in turn give large Bi_i and small η_i values. N = 3 is rarely needed. The number of elements Ne = 5–10 is usually sufficient for systems with non-stiff or slightly stiff concentration profiles. For very stiff cases, Ne = 20–30 is often enough.

Insufficient N tends to give diffused concentration profiles as shown in Figs. 3 to 5. Using N = 1 instead of N = 2 in Figs. 3 to 5 (dashed lines) saves about 60% CPU time on a SUN 4/280 computer, but the concentration profiles differ to some extent from the converged ones (solid lines). In Fig. 5, the dotted lines are obtained by using three quadratic finite elements (Ne = 3) and one interior collocation point (N = 1) with a CPU time of only 13.2 seconds. Though the dotted lines show a certain degree of oscillation, they still provide the general shapes of the converged concentration profiles, which take 6.9 minutes of CPU time. This means that one may use small Ne and N values to get the rough concentration profiles very quickly and then decide what to do next.

The efficiency and robustness of the numerical procedure are further demonstrated by more simulated effluent histories for the discussions in the following parts of this chapter, including cases involving very stiff concentration profiles.

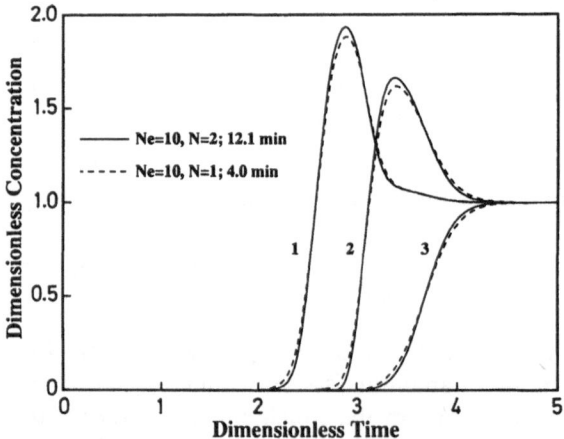

Fig. 3. Effect of the number of interior collocation points in the simulation of frontal adsorption

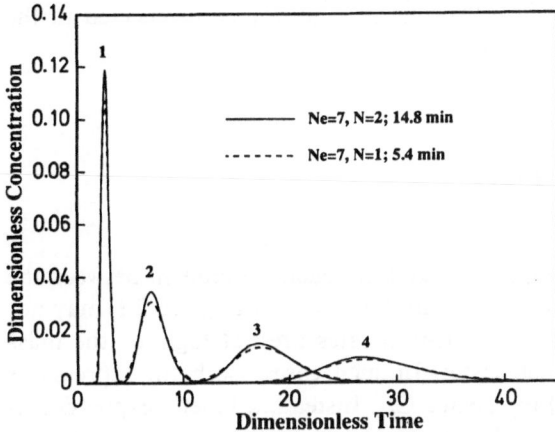

Fig. 4. Effect of the number of interior collocation points in the simulation of elution

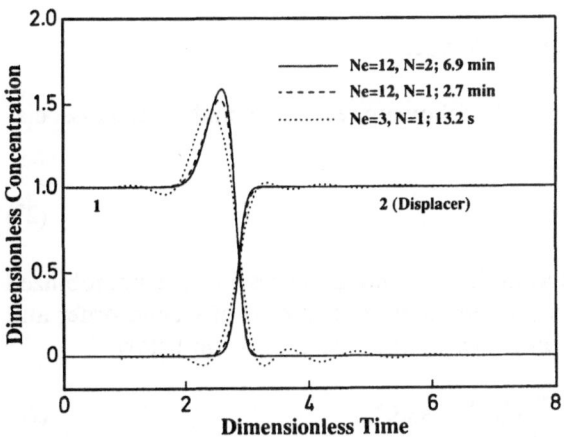

Fig. 5. Convergence of the concentration profiles of a stepwise displacement system

The FORTRAN code based on the numerical procedure discussed above is capable of simulating many kinds of multicomponent chromatographic processes, including frontal analysis, displacement development, simple nongradient elution, nonlinear gradient elution, and some multistage operations. Each mode of simulation is designated with a process index in the code, which is included in the data input.

The input data for the FORTRAN code contains the number of components, elements and interior collocation points, process index, time control data, dimensionless parameters, isotherm type and parameters. Note that the code is based on the dimensionless PDE systems and C_{0i} can be combined with b_i to form a dimensionless group $b_i C_{0i}$. The initial conditions are reflected in the process index, or entered in the data file.

5 Extension of the Rate Model

The assumption that a local equilibrium exists for each component between the pore surface and the stagnant fluid phase in the macropores (Sect. 2.1) may not be satisfied if the adsorption and desorption rates are not high, or the mass transfer rates are relatively much faster. In such cases, isotherm expressions cannot be inserted into Eq. (2) to replace C_{pi}^s. Instead, a kinetic expression is often used. The so-called second order kinetics has been widely used to account for reaction kinetics in the study of affinity chromatography [31–33, 35, 36, 38]. A general rate model with second order kinetics has been applied to affinity chromatography by Arve and Liapis [38].

5.1 Addition of Second Order Kinetics

The second order kinetics assumes the following reversible binding and dissocia-tion reaction.

$$P_i + L \underset{k_{di}}{\overset{k_{ai}}{\rightleftharpoons}} P_i L \tag{27}$$

where P_i is component i (macromolecule) and L represents the immobilized ligand. In this elementary reaction, the binding kinetics is of second order and the disassociation first order, as shown by the rate expression below.

$$\frac{\partial C_{pi}^s}{\partial t} = k_{ai} C_{pi} \left(C^\infty - \sum_{j=1}^{Ns} C_{pj}^s \right) - k_{di} C_{pi}^s \tag{28}$$

where k_{ai} and k_{di} are the adsorption and desorption rate constants for compo-nent i, respectively. The rate constant k_{ai} has a unit of concentration over time while the rate constant k_{di} has a unit of inverse time.

If the reaction rates are relatively large compared to mass transfer rates, then instant adsorption/desorption equilibrium can be assumed such that both sides of Eq. (28) can be set to zero, which consequently gives the Langmuir isotherms with the equilibrium constant $b_i = k_{ai}/k_{di}$ for each component.

Introducing dimensionless groups $Da_i^a = L\,(k_{ai} C_{0i})/v$ and $Da_i^d = Lk_{di}/v$ which are defined as the Damköhler numbers [54] for adsorption and desorp-tion, respectively, Eq. (28) can be nondimensionalized as follows.

$$\frac{\partial c_{pi}^s}{\partial \tau} = Da_i^a c_{pi} \left(c^\infty - \sum_{j=1}^{Ns} \frac{C_{0j}}{C_{0i}} c_{pj}^s \right) - Da_i^d c_{pi}^s \tag{29}$$

If the saturation capacities are the same for all the components, at equilibrium, Eq. (29) gives $b_i C_{0i} = Da_i^a/Da_i^d$ and $a_i = C^\infty b_i = c^\infty Da_i^a/Da_i^d$ for the resultant multicomponent Langmuir isotherm. The Damköhler numbers reflect the characteristic reaction times with that of the stoichiometric time. The $b_i = k_{ai}/k_{di}$ values for affinity chromatography are often very large [32], but it is erroneous to jump to the conclusion depending on this alone that the desorption rate must be much smaller than that of adsorption, since the two processes have different reaction orders, and the concentration C_{pi} is often very small on the adsorption side as expressed by the first term on the right hand side of Eq. (28). It is obvious that the dimensionless Damköhler numbers provide a better comparison in this regard.

5.2 Solution Strategy

Adding the second order kinetics to the general rate model does not complicate the numerical procedure for its solution since the discretization process is untouched. One only has to add Eq. (29) in the final ODE system.

The following equation should be used to replace Eq. (21)

$$(1 - \varepsilon_p)\frac{\partial c_{pi}^s}{\partial \tau} + \varepsilon_p \frac{\partial c_{pi}}{\partial \tau} = \eta_i \sum_{k=1}^{N+1} B_{1,k}(c_{pi})_k, l = 1, 2, \ldots, N \qquad (30)$$

The final ODE system consists of Eqs. (17), (29) and (30). With the trial values of c_{bi}, c_{pi} and c_{pi}^s in the function subroutine in the FORTRAN code, their derivatives can be easily evaluated from the three ODE expressions.

If Ne elements and N interior collocation points are used for the discretization of the Eqs. (1) and (2), there will be Ns (2Ne + 1) (2N + 1) ODE's in the final ODE system, which are Ns (2Ne + 1) N more than in the equilibrium case. These extra ODE's come from Eq. (29) at each element node and each interior collocation point for each component.

The relationship among the kinetic effects, reaction equilibrium and mass transfer rates were discussed by Gu [72].

5.3 Addition of Size Exclusion Effects

In some chromatographic systems, large solute molecules have considerable size exclusion effects, which means that such large molecules either cannot access part of the small macropores in the particles or the entire particle at all. This is especially true in affinity chromatography in which large macromolecules are often present, and sometimes even larger complexes can be formed between the macromolecules with the soluble ligands. Size exclusion effects reduces the saturation capacity of a component with a large molecular size. A new isotherm system was developed recently by Gu et al. [55], for the study of adsorption systems with uneven saturation capacities as a result of size exclusion.

In recent years, there have been three ACS Symposium Series on size exclusion chromatography [56, 57, 58]. Several mathematical models have been proposed for size exclusion chromatography [59, 60, 61] among which the model proposed by Kim and Johnson is particularly helpful for this work. Their model is similar to the general rate model described in Sect. 1.2 of this work, except that their model considers size exclusion single component systems involving no adsorption. They introduced an "accessible pore volume fraction" to account for the size exclusion effect.

In this work, the symbol ε_{pi}^a is used to denote the accessible porosity (i.e., accessible macropore volume fraction) for component i. It implies that for small molecules with no size exclusion effect, $\varepsilon_{pi}^a = \varepsilon_p$, and for large molecules that are completely excluded from the particles $\varepsilon_{pi}^a = 0$. For any medium-sized molecules $0 < \varepsilon_{pi}^a < \varepsilon_p$. It is convenient to define a size exclusion factor $0 \leq F_i^{ex} \leq 1$ such that $\varepsilon_{pi}^a = F_i^{ex}\varepsilon_p$. F_i^{ex} is a function of the distribution coefficient of component i. It is also a function of the particle size distribution if the particle sizes cannot be assumed to be equal [60]. To include the size exclusion effect, Eq. (2) should be modified as follows.

$$(1 - \varepsilon_p)\frac{\partial C_{pi}^s}{\partial t} + \varepsilon_{pi}^a \frac{\partial C_{pi}}{\partial t} - \varepsilon_{pi}^a D_{pi}\left[\frac{1}{R^2}\frac{\partial}{\partial R}\left(R^2\frac{\partial C_{pi}}{\partial R}\right)\right] = 0 \qquad (31)$$

where the first term $(1 - \varepsilon_p)\dfrac{\partial C_{pi}^s}{\partial t}$ should be dropped or set to zero, if a component does not bind with the stationary phase. It should be pointed out again that in the equation above C_{pi}^s values are based on the unit volume of the solids of the particles excluding the pores measured by the *particle porosity* ε_p. For a component which is completely excluded from the particles (i.e., $\varepsilon_{pi}^a = 0$), adsorbing only on the outer surface of the particles, Eq. (31) degenerates into the following interfacial mass balance relationship.

$$\frac{\partial C_{pi}^s}{\partial t} = \frac{3k_i}{(1 - \varepsilon_p)R_p}(C_{bi} - C_{pi, R=R_p}) \qquad (32)$$

This equation can be combined with the bulk-fluid phase governing equation (Eq. (1)) to give the following equation which is similar to a lumped particle model.

$$-D_{bi}\frac{\partial^2 C_{bi}}{\partial Z^2} + v\frac{\partial C_{bi}}{\partial Z} + \frac{\partial C_{bi}}{\partial t} + \frac{(1 - \varepsilon_b)(1 - \varepsilon_p)}{\varepsilon_b}\frac{\partial C_{pi}^s}{\partial t} = 0 \qquad (33)$$

where C_{pi}^s either follows the multicomponent isotherms or the expression for reversible binding, Eq. (28). If component i does not bind with the stationary phase, $C_{pi}^s \equiv 0$ and the fourth term in Eq. (33) is dropped for that component. As a reminder again, the solid phase concentration of component i, C_{pi}^s, is based on the unit volume of the solid part of the particle excluding pores, i.e., the unit volume of the solid skeleton. The dimensional form of Eq. (33) is listed below.

$$-\frac{1}{Pe_{Li}}\frac{\partial^2 c_{bi}}{\partial z^2} + \frac{\partial c_{bi}}{\partial z} + \frac{\partial c_{bi}}{\partial \tau} + \frac{(1-\varepsilon_b)(1-\varepsilon_p)}{\varepsilon_b}\frac{\partial c_{pi}^s}{\partial \tau} = 0 \qquad (34)$$

5.4 Solution Strategy

If no component is totally excluded, the addition of the size exclusion effect in the rate models is very simple. One only has to use $\varepsilon_{pi}^a D_{pi}$ to replace $\varepsilon_p D_{pi}$ in the expressions of Bi_i and η_i, and to use $\varepsilon_{pi}^a c_{pi}$ to replace $\varepsilon_p c_{pi}$ in Eq. (10).

Mathematically, a singularity occurs in the model equation system when a component (say, component i) is totally excluded from the particles (i.e., $\varepsilon_{pi}^a = 0$) if one does not use Eq. (34) to replace Eqs. (9) and (10). It turns out that for numerical calculation, there is no need to worry about this singularity, if ε_{pi}^a is given a very small value below that of the tolerance of the ODE solver, which is set to 10^{-5} throughout this work. It is found that this treatment gives the results which have the same values for the first five significant digits as those obtained by using Eq. (34).

One should be aware that the size exclusion of a component affects its saturation capacity in the isotherm. It also affects the effective diffusivity of the component since the tortuosity is related to accessible porosity. It is clear that using size exclusion in a multicomponent model often leads to the use of uneven saturation capacities for a component with significant size exclusion and a component without size exclusion. This may cause problems when the multicomponent Langmuir isotherm is used in terms of thermodynamic inconsistency [55].

6 Other Extensions of the Rate Model

The general rate model can also be modified to account for the interaction between adsorbates and soluble ligands as in affinity chromatography. This extension is considerably more complicated. Details were given by Gu [72].

7 Study of Stepwise Displacement

Stepwise displacement chromatography has received considerable attention recently in microbiological processes for in situ removal of toxic product(s) [62, 63, 64]. Lee et al. [30], used a polyvinylpyridine (PVP) resin for the in situ removal of lactic acid during growth. This kind of in situ separation reduces product inhibition, and thus enhances productivity. In such adsorption-combined processes, chromatographic columns are coupled with the bioreactor to remove the product(s) simultaneously via preferential adsorption and the adsorbate(s) is(are) then recovered through a displacement operation. This kind of stepwise displacement is also widely used to recover biomolecules from a dilute solution after they are adsorbed onto a column. In both cases, frontal adsorption proceeds the displacement process which often concentrates the adsorbate(s) by using a suitable displacer. This kind of stepwise displacement operation is somewhat different from the classical displacement chromatography or displacement development first classified by Tiselius [65] and extensively reviewed by Horvath and co-workers [11, 49, 66, 67].

Classical displacement chromatography was described by many researchers as a process in which a column packed with solid adsorbent is equilibrated with the mobile phase that has no or weak affinity to the adsorbent. A sample of mixtures is then introduced to the column. The sample usually takes up a fraction of the column volume in the inlet section. Subsequently, a development agent (called displacer) is pumped into the column. The displacer must have a higher affinity to the stationary phase than any of the components in the sample, i.e., its adsorption isotherm overlies those of the feed components [1, 67]. Provided that the column is sufficiently long, and isotherm curves are all favorably shaped, sample components will eventually migrate inside the column with the same speeds to form individual product zones. The series of such zones is usually called a displacement train [49, 66, 68, 69]. Figure 6 shows a displacement chromatogram with two sample solutes and one displacer. Parameter values used in simulation are listed in Table 1. Compared with elution chromatography, the displacement development has two distinct advantages: (1) the displacement effect reduces tailing (Fig. 6); (2) sample loading can be higher

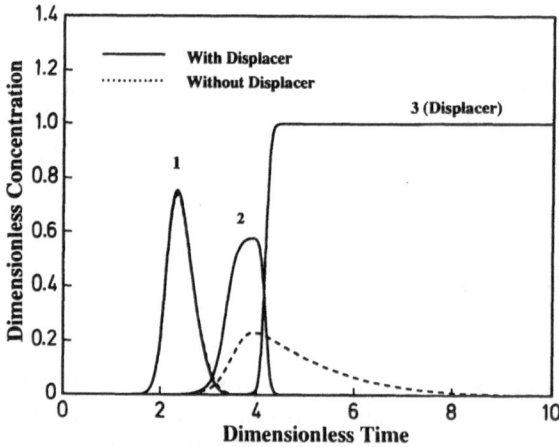

Fig. 6. Displacement chromatogram

Table 1. Parameter values used for simulation

Figures	Species	Physical Parameters					Numerical Parameters	
		Pe_{Li}	η_i	Bi_i	a_i	$b_i \times C_{0i}$	Ne	N
3	1	400	6	10	2	4×0.1	10	2
	2	400	6	10	7	14×0.1		
	3	400	6	10	15	30×0.1		
4	1	300	4	20	1.2	1.5×0.1	7	2
	2	320	4.2	17	8	10×0.1		
	3	400	5.5	16	24	30×0.1		
	4	500	7	15	38	48×0.1		
5	1	600	6	5	3	6×0.1	12	2
	2	600	3	6	12	24×0.3		
6	1	200	10	4	1	4×0.1	17	2
	2	200	5	6	5	20×0.1		
	3	400	5	6	30	120×0.1		
7	1	300	10	4	3	6×0.1	10	2
	2	300	15	4	4	8×0.3		
8	1	300	10	3	3	6×0.1	10	2
	2	300	15	4	4	8×0.11		
9	1	300	10	3	3	6×0.1	10	2
	2	300	15	4	2	4×0.4		
10	1	300	10	3	3	6×0.1	24	2
	2	300	15	4	30	60×0.3		

In all cases, $\varepsilon_b = 0.4$ and $\varepsilon_p = 0.5$. For Figure 4, $\tau_{imp} = 0.1$. The error tolerance of the ODE solver is tol $= 10^{-5}$. Double precision is used in the Fortran code

[11]. These features make the displacement development operation a very attractive alternative to elution in preparative scale chromatography [67].

The main difference between the displacement development and the stepwise displacement studied in this chapter is often the operation purpose itself. The former desires the products to be separated into a displacement train containing individual product zones in the effluent stream, while the latter requires the efficient displacement of the adsorbates. In other words, the desorption chromatography does not require a well-defined displacement train in the effluent; rather it requires the displacement of adsorbed component(s) with a minimum amount of displacer in a minimum length of time in order to obtain concentrated product(s). The product(s) in the effluent after displacement may be further purified if necessary after the stepwise displacement. A typical use of the stepwise displacement, as we have already mentioned, is the in situ separation during product formation [30, 62–64]. Another important difference is that the displacement development takes a sample which usually occupies only a fraction of the column inlet section while the stepwise displacement has no such limitation. The strong affinity of the displacer which is required in the displacement development should not be mistaken as a requirement for the stepwise displacement.

7.1 Results and Discussion [70]

The multicomponent Langmuir isotherms (Eq. (25)) with uniform adsorption saturation capacities will be used in this study. For simplicity, only the two component displacement chromatographic processes will be discussed, in which component 1 is the component to be displaced and component 2 the displacer.

7.2 Effect of Feed Concentration of Displacer (C_{02})

Figure 7 shows that the higher the displacer concentration in the mobile phase, the higher the roll-up peak on the concentration profile of component 1 (see Table 1 for parameter values). This is due to the fact that a higher displacer concentration in the feed gives a faster migration rate for the concentration front of the displacer inside the column, a larger $b_2 C_{p2}$ value and, hence, a better displacement efficiency. Figure 8 shows a case in which the displacer does not give much help in the displacement of component 1 from the column because the concentration of the displacer is too low. This kind of situation was mentioned by some researchers [5, 71]. In Fig. 9, the affinity of the displacer is lower than that of the adsorbate. It shows that if the concentration of the displacer is sufficiently high a desirable displacement of the adsorbate can also be achieved. The study of the effect of ethanol concentration on the efficiency of the displace-

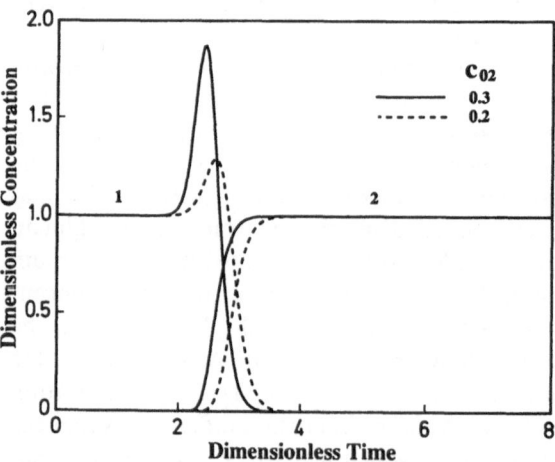

Fig. 7. Effect of displacer concentration on displacement

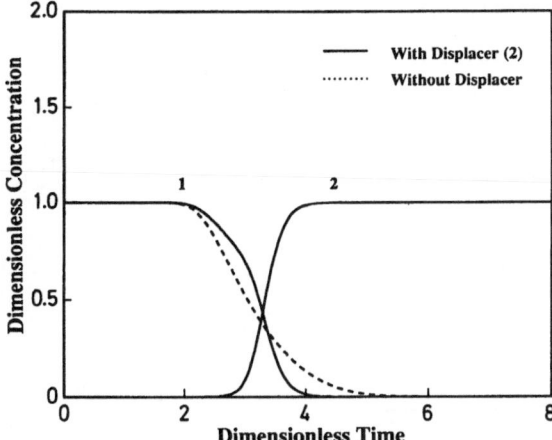

Fig. 8. Same conditions as Fig. 7, except that the concentration of the displacer is lower

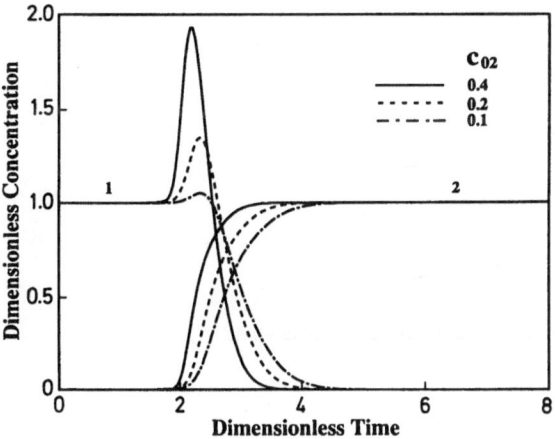

Fig. 9. Effect of displacer concentration on displacement for a case in which $b_2 < b_1$

ment of phenylalanine from a column packed with β-cyclodextrin-containing resin presented in Fig. 10 [72] qualitatively proved this argument. The affinity of ethanol with β-cyclodextrin is much smaller that phenylalanine, but when the ethanol concentration is sufficiently high, it still serves as an efficient displacer which gives good displacement results (Fig. 10).

7.3 Effect of Adsorption Equilibrium Constant of Displacer (b_2)

The effect of the b_2 value on displacement performance is shown in Fig. 11. It can be seen that an increase in b_2 delays the appearance of the roll-up peak,

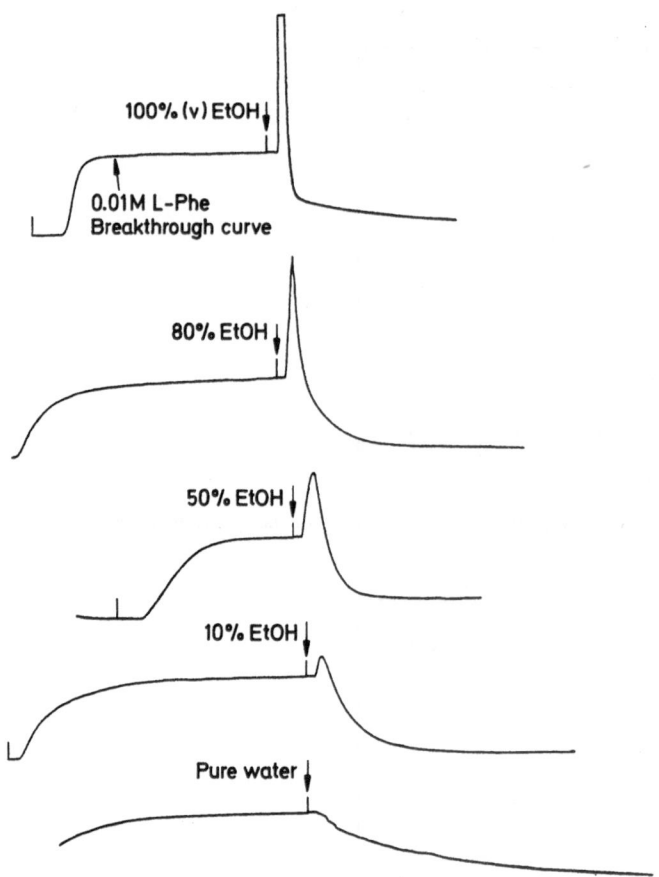

Fig. 10. Effect of ethanol concentration on displacement of phenylalanine

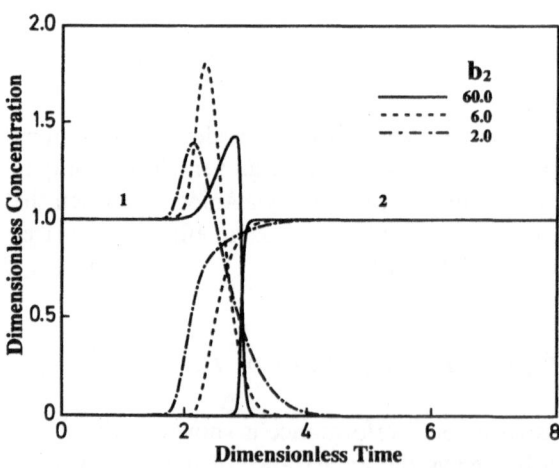

Fig. 11. Effect of b_2 on displacement performance

gives a sharper displacer front, and reduces the tailing of the displaced component. The maximum roll-up peak occurs somewhere in the middle range of the b_2 value. If the primary goal of displacement is to obtain a large fraction of pure component 1, a larger b_2 is obviously favorable. However, if the mixing of displacer in the product is not a setback, such as in the case when the displacer is a volatile organic solvent and the product is readily recovered by evaporation after the displacement, a larger b_2 is not always favorable. As a matter of fact, if the displacement is terminated when the major portion of the product has been recovered, then a small b_2 may be a better choice because the roll-up peak appears earlier.

Compared with the displacement development, the conclusion for stepwise displacement is somewhat different. The displacement development requires a displacer which has an affinity higher than any other component in the sample. However, this is hardly true for the stepwise displacement as we have already discussed in the cases of Figs. 9 to 11.

8 Summary

Among all kinds of models for nonlinear multicomponent chromatography, the general multicomponent rate model is the most comprehensive one. It accounts for various mass transfer mechanisms and nonlinear isotherms. It is a very useful tool for the study of the dynamics of nonlinear multicomponent chromatography. This chapter has presented an efficient numerical method for the solution and the extensions of the model. The model was used for the study of some interesting effects of isotherm characteristics of the displacer on the optimization of stepwise displacement. It was concluded that a displacer with a high feed concentration, and a suitable adsorption equilibrium constant is often a desirable choice when the purpose of the displacement operation is to displace and to concentrate the adsorbed species and to minimize the amount of displacer employed.

9 References

1. Ruthven D (1984) Principles of adsorption and adsorption processes. Wiley, New York
2. Glueckauf E (1949) Discuss Faraday Soc 7: 12
3. Helfferich F, Klein G (1970) Multicomponent chromatography theory of interference. Marcel Dekker, New York
4. Rhee H-K, Aris R, Amundson NR (1970) Philos Trans R Soc (London) Ser A 267: 419
5. Rhee H-K, Amundson NR (1982) AIChE J 28: 423
6. Lee CK, Yu Q, Kim SU, Wang N-HL (1989) J Chromatogr 484: 25
7. Glueckauf E (1947) J Chem Soc 1302

8. Helfferich F, James DB (1970) J Chromatogr 46: 1
9. Bailly M, Tondeur D (1981) Chem Eng Sci 36: 455
10. Frenz J, Horvath C (1985) AIChE J 31: 400
11. Frenz J, Horvath C (1988) High Performance Liquid Chromatography 5: 211
12. Yu Q, Yang J, Wang N-HL (1987) Reactive Polymers 6: 33
13. Wankat PC (1986) Large-scale adsorption and chromatography, vol 1. CRC Press, Boca Raton, FL
14. Martin AJP, Synge RLM (1941) Biochem J 35: 1359
15. Yang CM (1980) PhD Thesis, Purdue University, West Lafayette, IN
16. Villermaux J (1981) In: Rodrigues AE, Tondeur D (eds) Percolation processes: Theory and applications. Sijthoff and Noordhoff, Rockville, MD
17. Eble JE, Grob RL, Antle PE, Snyder LR (1987) J Chromatogr 405: 1
18. Seshadri S, Deming SN (1984) Anal Chem 56: 1567
19. Solms DJ, Smuts TW, Pretorius V (1971) J Chromatogr Sci 9: 600
20. Eble JE, Grobe RL, Antle PE, Snyder LR (1987) J Chromatogr 384: 25
21. Glueckauf E, Coates JI (1947) J Chem Soc 1315
22. Rhee H-K, Amundson NR (1974) Chem Eng Sci 29: 2049
23. Bradley WG, Sweed NH (1975) AIChE Symp Ser 71: 59
24. Golshan-Shirazi S, Guiochon G (1988) J Chromatogr 461: 1
25. Golshan-Shirazi S, Guiochon G (1988) J Chromatogr 461: 19
26. Farooq S, Ruthven DM (1990)
27. Zwiebel I, Gariepy RL, Schnitzer JJ (1972) AIChE J 18: 1139
28. Santacesaria E, Morbidelli M, Servida A, Storti G, Carra S (1982) Ind Eng Chem Process Des Dev 21: 446
29. Santacesaria E, Morbidelli M, Servida A, Storti G, Carra S (1982) Ind Eng Chem Process Des Dev 21: 451
30. Lee S, Tsai G-J, Seo JH, Tsao GT (1988) Third Chemical Congress Of North America and 195th ACS National Meeting. Toronto
31. Chase HA (1984) Chem Eng Sci 39: 1099
32. Chase HA (1984) J Chromatogr 279: 179
33. Arnold FH, Blanch HW, Wilke CR (1985) J Chromatogr 30: B9
34. Arnold FH, Blanch HW, Wilke CR (1985) J Chromatogr 30: B25
35. Arnold FH, Schofield SA, Blanch HW (1986) J Chromatogr 355: 1
36. Arnold FH, Schofield SA, Blanch HW (1986) J Chromatogr 355: 13
37. Arve BH, Liapis AI (1987) AIChE J 33: 179
38. Arve BH, Liapis AI (1987) Biotechnol Bioeng 30: 638
39. Arve BH, Liapis AI (1988) Biotechnol Bioeng 31: 240
40. Lee W-C (1989) PhD Thesis, Purdue University, West Lafayette, IN
41. Ramkrishna D, Amundson NR (1985) Linear operator methods in chemical engineering with applications to transport and chemical reaction systems. Prentice-Hall, Englewood Cliffs, NJ
42. Gu T, Tsai G-J, Tsao GT (1990) AIChE J 36: 784
43. Liapis AI, Rippin DWT (1978) Chem Eng Sci 33: 593
44. Yu Q, Wang N-HL (1989) Computers Chem Eng 13: 915
45. Mansour A (1989) Sep Sci Technol 24: 1047
46. Mansour A, von Rosenberg DU, Sylvester ND (1982) AIChE J 28: 765
47. Villadsen J, Michelsen ML (1978) Solutions of differential equation models by polynomial approximation. Prentice Hall, Englewood Cliffs, NJ
48. Finlayson BA (1980) Nonlinear analysis in chemical engineering. McGraw-Hill, New York
49. Antia FD, Horvath C (1989) Ber Busenges Phys Chem 93: 961
50. IMSL (1987) IMSL User's Manual, Version 1.0. IMSL, Inc. Houston 640–652
51. Gear C (1972) Numerical initial-value problems in ordinary differential equations. Prentice-Hall, Englewood Cliffs, NJ
52. Gardini L, Servila A, Morbidelli M, Carra S (1985) Computers Chem Eng 25: 490
53. Reddy JN (1984) An introduction to the finite element method. McGraw Hill, New York
54. Froment GF, Bischoff KB (1979) Chemical reactor analysis and design. Wiley, New York
55. Gu T, Tsai G-J, Tsao GT (1991) AIChE J 37: 1333
56. Provder T (ed) (1980) ACS Symp Series. No 138
57. Provder T (ed) (1984) ACS Symp Series. No 245
58. Provder T (ed) (1984) ACS Symp Series. No 352

59. Yau WW, Kirkland JJ, Bly DD (1979) Modern size-exclusion liquid chromatography, Wiley, New York
60. Kim DH, Johnson AF (1984) In: Provder T (ed) ACS Symposium Series 245: 25
61. Koo Y-M, Wankat PC (1988) Sep Sci Technol 23: 413
62. Yang X (1988) M.S. Thesis, Purdue University, West Lafayette, IN
63. Yang X, Tsai G-J, Tsao GT (1988) Third Chemical Congress Of North America and 195th ACS National Meeting, Toronto
64. Yang X, Tsai G-J, Tsao GT (1989) AIChE Summer National Meeting, Philadelphia
65. Tiselius A (1940) Ark Kem Mineral Geol 14B: 1
66. Horvath C, Nahum A, Frenz JH (1981) J Chromatogr 218: 365
67. Horvath C (1985) In: Bruner F (ed) The Science of Chromatography. Elsevier, New York
68. Phillips WM, Cramer SM (1988) J Chromatogr 454: 1
69. Katti AM, Guichon GA (1988) J Chromatogr 449: 25
70. Gu T, Tsai G-J, Tsao GT (1991) Biotechnol Bioeng 37: 65
71. Morbidelli M, Storti G, Carra S, Niederjaufner G, Pontoglio A (1985) Chem Eng Sci 40: 1155
72. Gu T (1990) PhD Thesis, Purdue University, West Lafayette, IN

Multicomponent Radial Flow Chromatography

T. Gu[1], G-J. Tsai[2], and G. T. Tsao[3]
[1] Department of Chemical Engineering, Ohio University, Athens, Ohio 45701, USA
[2] Building 130, Lederle Laboratories, Pearl River, NY 10965, USA
[3] Laboratory of Renewable Resources Engineering, 1295 Potter Center, Purdue University, West Lafayette, IN 479907-1295, USA

Radial flow chromatography (RFC) is an alternative to the conventional axial flow chromatography (AFC) in preparative and large scale applications. It provides a faster flow rate and a lower bed pressure drop. In this chapter a general nonlinear multicomponent rate model for RFC is presented. The radial dispersion and mass transfer coefficients are treated as variables in the model. The model was solved numerically by using finite element and orthogonal collocation methods for the discretizations of bulk-fluid and particle phase partial differential equations, respectively. Simulated examples are given for various chromatographic operations. The diffusional and mass transfer effects in RFC are studied. Some comparisons between AFC and RFC are shown. The model is also extended to include second order kinetics, size exclusion effect and reaction in the liquid phase for modeling of biospecific elution using soluble ligands.

Advances in Biochemical Engineering
Biotechnology, Vol. 49
Managing Editor: A. Fiechter
© Springer-Verlag Berlin Heidelberg 1993

List of Symbols and Abbreviations

Symbol	Description
a_i	constant in Langmuir isotherm for component i, $b_i\,C_i^\infty$
b_i	adsorption equilibrium constant for component i, k_{ai}/k_{di}
Bi_i	Biot number of mass transfer for component i, $k_i R_p/(\varepsilon_p D_{pi})$
$\overline{Bi}_i$	averaged Bi_i
C_{bi}	bulk-fluid phase concentration of component i
C_{fi}	feed concentration profile of component i, a time dependent variable
C_{0i}	concentration used for nondimensionalization, $\max\{C_{fi}(t)\}$
C_{pi}	concentration of component i in the stagnant fluid phase inside particle macropores
C_{pi}^s	concentration of component i in the solid phase of particle (mole absorbate/unit volume of particle skeleton)
C_i^∞	adsorption saturation capacity for component i (mole adsorbate/ unit volume of particle skeleton)
c_{bi}	$= C_{bi}/C_{0i}$
c_{pi}	$= C_{pi}/C_{0i}$
c_{pi}^s	$= C_{pi}^s/C_{0i}$
c_i^∞	$= C_i^\infty/C_{0i}$
D_{bi}	axial or radial dispersion coefficient of component i
$\overline{D}_{bi}$	averaged D_{bi}
D_{pi}	effective diffusivity of component i, porosity not included
h	axial bed length of the radial flow column
k_i	film mass transfer coefficient of component i
k_{ai}	adsorption rate constant for component i
k_{di}	desorption rate constant for component i
N	number of interior collocation points
Ne	number of quadratic elements
Ns	number of components
Pe_i	Peclet number of radial dispersion for component i, $\dfrac{v(X_1 - X_0)}{D_{bi}}$
Q	volumetric flow rate of the mobile phase
R	radial coordinate for particle
R_p	particle radius
r	$= R/R_p$
t	time
v	interstitial velocity
V_b	bed volume for RFC column, $\pi h(X_1^2 - X_0^2)$
V_0	dimensionless constant, $\dfrac{\pi h X_0^2}{V_b}$ or $\dfrac{X_0^2}{X_1^2 - X_0^2}$

V	dimensionless volumetric coordinate, $\dfrac{\pi h(X^2 - X_0^2)}{V_b}$
	or $\dfrac{X^2 - X_0^2}{X_1^2 - X_0^2} \in [0, 1]$
X	radial coordinate for RFC column
X_1	outer radius of RFC column
X_0	inner radius of RFC column

Greek Letters

α	$= 2\sqrt{V + V_0}(\sqrt{1 + V_0} - \sqrt{V_0})$
ε_b	bed void volume fraction
ε_p	particle porosity
η_i	dimensionless constant, $\dfrac{\varepsilon_p D_{pi}}{R_p^2} \dfrac{V_b \varepsilon_b}{Q}$
ξ_i	dimensionless constant for component i, $3Bi_i \eta_i (1 - \varepsilon_b)/\varepsilon_b$
τ	dimensionless time, $\dfrac{Qt}{V_b \varepsilon_b}$
τ_{imp}	dimensionless time duration for a rectangular pulse of a sample
ϕ	Lagrangian interpolation function

1 Introduction

Chromatography has long been established as an effective means of separation. It is becoming more and more popular in the age of the rapid development of biotechnology. The demand for efficient preparative and large scale liquid chromatographic separation processes is steadily increasing. Radial flow chromatography (RFC), since its introduction onto the commercial market in the mid 1980s [1], has proved to be a promising alternative to conventional axial flow chromatography (AFC). Compared to AFC, the RFC geometry (Fig. 1) provides a relatively large flow area and short flow path. These factors enable a larger volumetric flow rate and shorter shift time in liquid chromatographic separations. Its soft gels or affinity matrix materials are used as separation media, the low pressure drop of RFC helps prevent bed compression [2, 3]. A full range of sizes from 50 ml to 200 l in bed volume of RFC columns both prepacked and unpacked can be obtained from commercial companies. Separation of various biological products has been reported [2, 4, 5, 6, 7, 8, 9]. An experimental case study of the comparison of RFC and AFC was carried out by Saxena and Weil [5] for the separation of ascites using QAE cellulose packings. They reported that by using a higher flow rate, the separation time for RFC was one-fourth that needed for a longer AFC column with the same bed volume. It was claimed that by using RFC instead of AFC, separation productivity can be improved quite significantly [1]. RFC is especially suitable for affinity chromatography in which solutes are usually strongly retained. This permits the use of a high flow

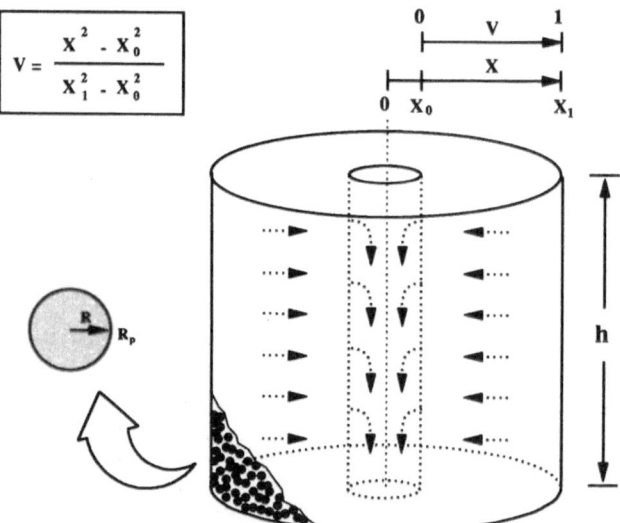

$$V = \frac{X^2 - X_0^2}{X_1^2 - X_0^2}$$

Fig. 1. Structure of a cylindrical radial flow column

rate and a short flow path for the fast treatment of a large volume of samples. RFC is advantageous for separation processes in which the effluent from the column is being recycled and the adsorbate will not be lost in the effluent stream [10], such as in the case of in situ separation (integrated process) [11].

Radial flow packed-bed reactors have been used for a variety of industrial applications such as ammonia and methanol syntheses, catalytic reforming, and vapor-phase desulfurization [12, 13]. Unlike the radial flow reactor, RFC is used primarily for liquid phase chromatographic separations. There are quite a few papers that are related to the theoretical and experimental studies of radial flow reactors. A brief review was given by Hlavacek and Votruba [14]. More recent papers include Strauss and Budde [13], Balakotaiah and Luss [12], Chang et al. [15], Lepez de Ramos and Pironti [16], and Tharakan and Chau [17]. The study of RFC may provide some useful information for the understanding of mass transfer processes in radial flow reactors.

Because of the special flow geometry in RFC some complications may arise in mathematical modeling. Since the linear flow velocity (v) in the RFC column changes constantly along the radial coordinate of the column (Fig. 1), unlike in most cases of AFC, the radial dispersion and external mass transfer coefficients are no longer constants. This important feature was rarely considered in the mathematical modeling of RFC in the literature in the past. Extensive theoretical studies have been reported for single component ideal RFC, which neglects radial dispersion, intraparticle diffusion, and external mass transfer resistance. In such studies a local equilibrium and a linear isotherm were often assumed. The earliest theoretical treatment of RFC was done by Lapidus and Amundson [18]. A similar study was carried out by Rachinskii [19]. Later Inchin and Rachinskii [20] included bulk-fluid phase molecular diffusion in their model. Lee et al. [21], proposed a unified approach for moments in linear chromatography, both AFC and RFC. They used several single component rate models for the comparison of statistical moments for RFC and AFC. Their models included radial dispersion, intraparticle diffusion, and external mass transfer effects. Kalinichev and Zolotarev [22] also carried out an analytical study on moments for single component RFC in which they treated the radial dispersion coefficient as a variable.

A rate model for nonlinear single component RFC was solved numerically by Lee [23] by using the finite difference and orthogonal collocation methods. His model considered radial dispersion, intraparticle diffusion, external mass transfer, and nonlinear isotherms. It used averaged radial dispersion and mass transfer coefficients instead of treating them as variables. A nonlinear model of this kind of complexity has no analytical solution and must be solved numerically.

Rhee et al. [24], discussed the extension of their multicomponent chromatography theory for ideal AFC with Langmuir isotherms, which is a parallel treatise to the interference theory developed by Helfferich [25], to RFC. Apart from this, so far no other detailed theoretical treatment of nonlinear multicomponent RFC is available in the literature. With the development of powerful computers and efficient numerical methods, more complicated treatment of multicomponent RFC now becomes possible. A general model for multicomponent RFC can provide some very valuable information.

In this chapter a numerical procedure is presented for solution to a general rate model for multicomponent RFC. The model is solved numerically. The solution of the model enables the discussion of several important issues concerning the characteristics and performance of RFC and its differences from AFC. And also the question of whether one should treat dispersion and mass transfer coefficients as variables in RFC.

2 General Multicomponent Rate Model for RFC

Consider a fixed bed with cylindrical radial flow geometry (Fig. 1) which is filled with uniform spherical porous solid adsorbents. Suppose the process is isothermal and there is no concentration gradient in the axial direction of the column. Although it can be a problem in some real cases, the possible maldistribution of flow streams is ignored in this theoretical study. Also, local equilibrium is assumed for each component between the pore surface and the liquid phase in the macropores in particles. Based on these basic assumptions, the following governing equations for component i in the bulk-fluid and particle phases via mass balances in the two phases can be formulated.

$$-\frac{1}{X}\frac{\partial}{\partial X}\left(D_{bi}X\frac{\partial C_{bi}}{\partial X}\right) \pm v\frac{\partial C_{bi}}{\partial X} + \frac{\partial C_{bi}}{\partial t}$$

$$+\frac{3k_i(1-\varepsilon_b)}{\varepsilon_b R_p}(C_{bi} - C_{pi, R=R_p}) = 0 \qquad (1)$$

$$\frac{\partial}{\partial t}[(1-\varepsilon_p)C_{pi}^s + \varepsilon_p C_{pi}] - \varepsilon_p D_{pi}\left[\frac{1}{R^2}\frac{\partial}{\partial R}\left(R^2\frac{\partial C_{pi}}{\partial R}\right)\right] = 0 \qquad (2)$$

where in Eq. (1) $+ v$ is for outward flow and $- v$ for inward flow. Note that in Eq. (1) D_{bi} and k_i are variables which are dependent on v.

The initial and boundary conditions are

$$t = 0, C_{bi} - C_{bi}(0, X) \quad C_{pi} = C_{pi}(0, R, X) \qquad (3, 4)$$

$$R = 0, \frac{\partial C_{pi}}{\partial R} = 0 \quad R = R_p, \frac{\partial C_{pi}}{\partial R} = \frac{k_i}{\varepsilon_p D_{pi}}(C_{bi} - C_{pi, R=R_p}) \qquad (5, 6)$$

For outward flow

$$X = X_0, \frac{\partial C_{bi}}{\partial X} = \frac{v}{D_{bi}}(C_{bi} - C_{fi}(t)) \quad X = X_1, \frac{\partial C_{bi}}{\partial X} = 0 \qquad (7a, 8a)$$

and for inward flow

$$X = X_1, \frac{\partial C_{bi}}{\partial X} = \frac{v}{D_{bi}}(C_{bi} - C_{fi}(t)) \quad X = X_0, \frac{\partial C_{bi}}{\partial X} = 0 \qquad (7b, 8b)$$

These equations can be expressed in the following dimensionless forms,

$$-\frac{\partial}{\partial V}\left(\frac{\alpha}{Pe_i}\frac{\partial c_{bi}}{\partial V}\right) \pm \frac{\partial c_{bi}}{\partial V} + \frac{\partial c_{bi}}{\partial \tau} + \xi_i(c_{bi} - c_{pi,r=1}) = 0 \qquad (9)$$

$$\frac{\partial}{\partial \tau}[(1 - \varepsilon_p)c_{pi}^s + \varepsilon_p c_{pi}] - \eta_i\left[\frac{1}{r^2}\frac{\partial}{\partial r}\left(r^2\frac{\partial c_{pi}}{\partial r}\right)\right] = 0 \qquad (10)$$

where for bulk-fluid phase equation (Eq. (9)) the local volume averaging method of Lee [23] and Slattery [26] has been used for its nondimensionalization. A comparison of the definitions of the dimensionless variables and parameters used in AFC and RFC is listed in Table 1.

In Eq. (9), the radial flow Peclet number is defined as $Pe_i = v(X_1 - X_0)/D_{bi}$. The introduction of the radial flow Peclet number and the use of the local volume averaging method in the transformation, streamline the analogy and

Table 1. Comparison of dimensionless variables and parameters

Symbols	AFC	RFC
τ	$\dfrac{t}{\left(\dfrac{L}{v}\right)}$	$\dfrac{t}{\left(\dfrac{V_b\varepsilon_b}{Q}\right)}$
z, V	$z = \dfrac{Z}{L}$	$V = \dfrac{X^2 - X_0^2}{X_1^2 - X_0^2}$
r	$\dfrac{R}{R_p}$	Same
Pe	$Pe_{Li} = \dfrac{vL}{D_{bi}}$	$Pe_i = \dfrac{v(X_1 - X_0)}{D_{bi}}$
Bi_i	$\dfrac{k_i R_p}{\varepsilon_{pi}^a D_{pi}}$	Same
η_i	$\dfrac{\varepsilon_{pi}^a D_{pi}}{R_p^2}\dfrac{L}{v}$	$\dfrac{\varepsilon_{pi}^a D_{pi}}{R_p^2}\dfrac{V_b\varepsilon_b}{Q}$
ξ_i	$\dfrac{3Bi_i\eta_i(1 - \varepsilon_b)}{\varepsilon_b}$	Same
V_0	—	$\dfrac{X_0^2}{X_1^2 - X_0^2}$
α	—	$2\sqrt{V + V_0}(\sqrt{1 + V_0} - \sqrt{V_0})$

comparison between the RFC model and the AFC model [27]. One may find that the dimensionless RFC rate model looks very similar to the dimensionless AFC rate model shown in an article by the same authors [28], except that in the RFC model there is an extra variable α and ξ is not a constant. Also, in RFC, there are two different flow directions, which are reflected by the $\pm$ sign in Eq. (9).

Initial conditions:

$$\tau = 0, \quad c_{bi} = c_{bi}(0, V), \quad c_{pi} = c_{pi}(0, r, V) \tag{11, 12}$$

Boundary conditions for outward flow:

$$V = 0, \frac{\partial c_{bi}}{\partial V} = \frac{Pe_i}{\alpha} \left[c_{bi} - \frac{C_{fi}(\tau)}{C_{0i}} \right] \tag{13}$$

For frontal adsorption, $C_{fi}(\tau)/C_{0i} = 1$

For elution, $C_{fi}(\tau)/C_{0i} = \begin{cases} 1 & 0 \leq \tau \leq \tau_{imp} \\ 0 & \text{else} \end{cases}$

After the sample has been introduced:

(in the form of frontal adsorption)

if component i is displaced, $\dfrac{C_{fi}(\tau)}{C_{0i}} = 0$

if component i is a displacer, $\dfrac{C_{fi}(\tau)}{C_{0i}} = 1$

$$V = 1, \quad \frac{\partial c_{bi}}{\partial V} = 0 \tag{14}$$

$$r = 0, \quad \frac{\partial c_{pi}}{\partial r} = 0 \quad r = 1, \frac{\partial c_{pi}}{\partial r} = Bi_i(c_{bi} - c_{pi, r=1}) \tag{15, 16}$$

For inward flow one only needs to swap $V = 0$ in Eq. (13) with $V = 1$ in Eq. (14).

The general solution strategy for the coupled partial differential equation (PDE) system (Eqs. (9) and (10)) is the same as that for the case of AFC discussed in the article by the same authors [28]. Compared to AFC, the solution for RFC seems to be more complicated because of variations in some physical properties of the system as mentioned earlier. The finite element method is apparently an ideal approach for this kind of system.

3 Numerical Solution

Equations (9) and (10) are transformed to a set of ODEs by the finite element method and the orthogonal collocation (OC) method [29], respectively.

Using the Galerkin approximation [30], Eq. (9) becomes

$$[DB_i][c'_{bi}] + [AKB_i][c_{bi}] = [PB_i] + [AFB_{bi}] \tag{17}$$

where $\quad (DB_i)^e_{m,n} = \int \phi_m \phi_n \, dV \tag{18}$

$$(AKB_i)^e_{m,n} = \int \left(\frac{\alpha}{Pe_i} \frac{\partial \phi_m}{\partial V} \frac{\partial \phi_n}{\partial V} \pm \phi_m \frac{\partial \phi_n}{\partial V} + \xi_i \phi_m \phi_n \right) dV \tag{19}$$

$$(AFB_i)^e_m = \int \xi_i \phi_m c_{pi, r=1} \, dV \tag{20}$$

in which $m, n = 1, 2, 3$, and the superscript e indicates that the finite element matrices and vectors are evaluated over each individual element before global assembly. In Eq. (19), $+ \phi_m$ is for outward flow and $- \phi_m$ for inward flow. The natural boundary condition $(PB_i) = - c_{bi} + C_{fi}(\tau)/C_{0i}$ is applied to $[AKB_i]$ and $[AFB_i]$ at $V = 0$ for outward flow or $(PB_i) = c_{bi} - C_{fi}(\tau)/C_{0i}$ at $V = 1$ for inward flow. $(PB_i) = 0$ elsewhere. Note that in Eq. (19) α is a function of V.

The particle phase equation can be discretized with N interior collocation points. The ODE system resulting from the numerical discretization can then be solved by Gear's stiff method. The procedure is the same as that for AFC which has been shown in the article by the same authors [28].

In this numerical procedure D_{bi} and k_i values are treated as variables which are dependent on the variations of v along the radial coordinate V. Meanwhile intraparticle diffusivities (D_{pi}) are regarded as independent of the variations of v. For dispersion coefficient, $D_b \propto v$ [23, 31]. This also applies to RFC. Thus the Pe_i can be treated as constants that are independent of v. It has also been shown [9] that

$$k_i \propto v^{1/3} \tag{21}$$

Since $k_i \propto v^{1/3}$ and

$$v \propto \frac{1}{X} \propto \frac{1}{\sqrt{V + V_0}} \tag{22}$$

one has

$$Bi_i \propto k_i \propto (1/X)^{1/3} \propto (V + V_0)^{-1/6} \tag{23}$$

If $Bi_{i, V=1}$ (i.e., $Bi_{i, X=X_1}$) values are given as input values, then

$$Bi_{i, V} = \left[\frac{1 + V_0}{V + V_0} \right]^{1/6} Bi_{i, V=1} \tag{24}$$

For ξ_i at any V position one has

$$\xi_{i, V} = \frac{3\eta_i(1 - \varepsilon_b) Bi_{i, V}}{\varepsilon_b} = \frac{3\eta_i(1 - \varepsilon_b)}{\varepsilon_b} \left[\frac{1 + V_0}{V + V_0} \right]^{1/6} Bi_{i, V=1} \tag{25}$$

Equation (25) is used for the evaluation of Eq. (19). In Eq. (19), due to the special geometry of radial flow chromatography, there are two space coordinates (V) dependent variables, α and ξ_i. The finite element integral in Eq. (19) is evaluated for each local element and α, ξ_i can be dealt with routinely without any trouble, since in this chapter finite element integrals are evaluated using four point Gauss-Legendre quadratures [29]. The ability to deal with variable physical properties with ease is one of the well-known advantages of the finite element method. Accuracy is another notable advantage of the method. The accommodation of variable Bi_i in Eq. (22) for the particle phase is also very easy. Since particle phase equations must be solved at each finite element node (with given nodal position, V) in the function subroutine, $Bi_{i,V}$ values can be readily obtained from Eq. (24).

It is very helpful to study the effects of treating D_{bi} and k_i as variables compared to treating them as constants, as is the case in most of the existing papers in the literature. There are two easy ways to take averaged D_{bi} and k_i values for the modeling. In both cases $Pe_i = v(X_1 - X_0)/D_{bi}$ will no longer be constant. In all the RFC cases in this chapter in the input data for simulation $Bi_{i,V=1}$ values are given.

The discussion below shows how to modify the algorithm to accommodate cases using the averaged D_{bi} and k_i values in order to make comparisons.

Using v value at V = 0.5 for averaging D_{bi} and k_i

$$\frac{\alpha}{\overline{Pe_i}} = \frac{\alpha}{\dfrac{v(X_1 - X_0)}{\overline{D}_{bi}}} = \frac{\alpha}{\dfrac{v(X_1 - X_0)}{D_{bi}}\dfrac{D_{bi}}{\overline{D}_{bi}}} = \frac{\alpha}{Pe_i}\frac{\overline{D}_{bi}}{D_{bi}} \tag{26}$$

Since $D_{bi} \propto v$, from Eq. (22) it is easy to obtain

$$\frac{\alpha}{\overline{Pe_i}} = \frac{\alpha}{Pe_i}\sqrt{\frac{V + V_0}{0.5 + V_0}} \tag{27}$$

Eq. (24) gives

$$\overline{Bi}_i = Bi_{i,V=0.5} = \left[\frac{1 + V_0}{0.5 + V_0}\right]^{1/6} Bi_{i,V=1} \tag{28}$$

Using v value at X = (X_1 + X_0)/2 for averaging D_{bi} and k_i

$$\frac{\alpha}{\overline{Pe_i}} = \frac{\alpha}{Pe_i}\frac{\overline{D}_{bi}}{D_{bi}} = \frac{\alpha}{Pe_i}\frac{X}{(X_1 + X_0)/2} = \frac{\alpha}{Pe_i}\frac{\sqrt{V + V_0}}{(\sqrt{1 + V_0} + \sqrt{V_0})/2} \tag{29}$$

For Bi_i one has

$$\frac{Bi_{i(X_1 + X_0)/2}}{Bi_{i,X_1}} = \left[\frac{X_1}{(X_1 + X_0)/2}\right]^{1/3} = \left[\frac{\sqrt{1 + V_0}}{(\sqrt{1 + V_0} + \sqrt{V_0})/2}\right]^{1/3} \tag{30}$$

which gives

$$\overline{Bi}_i = Bi_{i,(X_1 + X_0)/2} = \left[\frac{\sqrt{1 + V_0}}{(\sqrt{1 + V_0} + \sqrt{V_0})/2}\right]^{1/3} Bi_{i,V=1} \tag{31}$$

If using averaged values of D_{bi} and k_i instead of treating them as v dependent variables, one needs to use $\alpha/\overline{Pe_i}$ in Eq. (27) or Eq. (29) to replace α/Pe_i in Eq. (19), and use $\overline{Bi_i}$ to replace Bi_i in Eq. (23). Bi_i should also be used to calculate ξ_i.

4 Results and Discussion

The multicomponent Langmuir isotherm and the stoichiometric isotherm are used for the simulations. Parameter values used in simulations are listed in Table 2, or mentioned during discussions.

4.1 Simulations of Different Chromatographic Operations

Figure 2 shows the simulated breakthrough curves for two components in inward and outward flow RFC. The corresponding breakthrough curves in AFC are also shown for comparison. They were obtained by using the same dimensionless parameters and isotherms as in the RFC case, except that $Bi_1 = Bi_2 = 11.15$ were used. These two Bi values for AFC are the ones its corresponding RFC possesses at $V = 0.5$. Figure 2 clearly shows that in RFC, inward flow provides sharper concentration profiles than outward flow. This is in agreement with the results obtained by Lee [23] for single component RFC. For a comparison similar to that shown in Fig. 2, the simulated effluent history

Table 2. Parameter values used for simulation*

| Figures | Species | Physical parameters | | | | | Numerical parameters | |
		Pe_{Li}	η_i	Bi_i	a_i	$b_i \times C_{0i}$	Ne	N
2	1	30	8	10	4	5×0.2	8	2
	2	30	8	10	8	10×0.2		
3	1	25	10	3	4	5×0.1	8	2
	2	25	10	3	8	10×0.2		
4	1	50	15	12	1	2×0.2	6	2
	2	65	20	8	10	20×0.2		
	1	90	10	40				
5	2	90	10	40			5	2
	3	90	10	40				
6	1	30	10	20	5	5×0.2	8	2
	2	30	10	20	6	6×0.8		

* In all runs, $\varepsilon_p = \varepsilon_b = 0.4$, and $V_0 = 0.04$ or otherwise as specified. For elution cases, sample size is $\tau_{imp} = 0.2$. For Fig. 5, $\alpha_{13} = 0.6$, $\alpha_{23} = 3$, $C_{01} = C_{02} = 0.1$, $C_{03} = 0.04$ N and $\bar{C} = 1.8$ meq ml^{-1} bed. The ODE solver's (IMSL's IVPAG subroutine) tolerance is tol $= 10^{-5}$. Double precision is used in the Fortran code. CPU times on SUN 4/280 computer for some simulations are: Fig. 2 (solid lines), 4.7 min; Fig. 3 (solid lines), 0.9 min; Fig. 4 (solid lines), 4.5 min

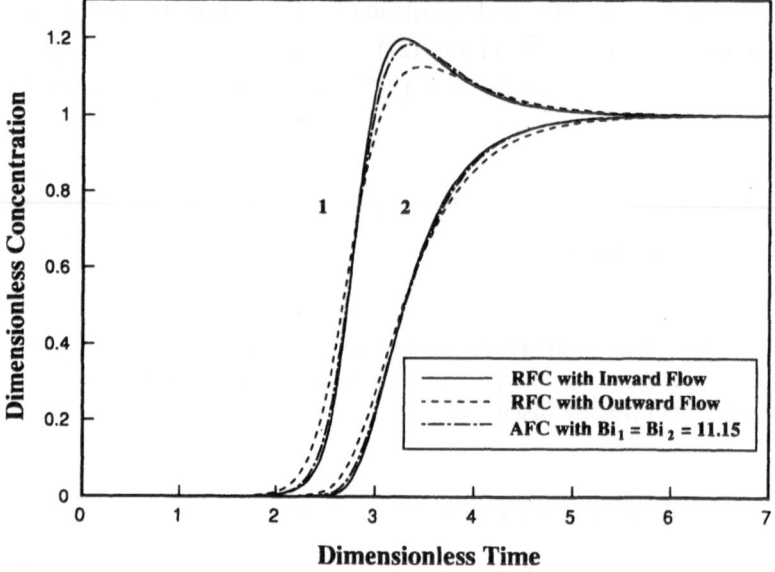

Fig. 2. Comparison of inward and outward flow RFC and AFC in frontal adsorption

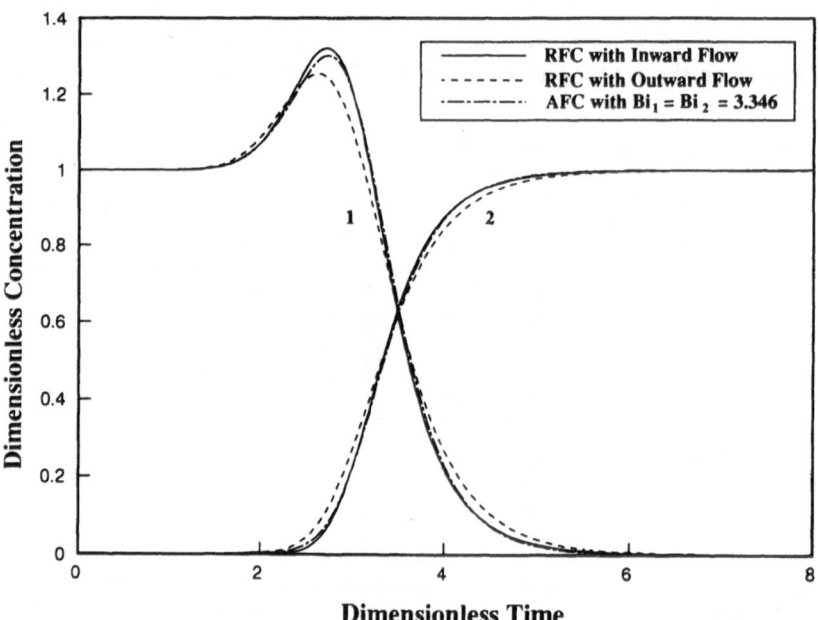

Fig. 3. Comparison of inward and outward flow RFC and AFC in displacement

for a step change displacement process is presented in Fig. 3. In this case the column is presaturated with component 1. Component 2 is introduced via a step change as a displacer to displace the adsorbed molecules of component 1. Again, inward flow RFC offers sharper concentration profiles, which are favorable for separation. The same conclusion also holds for the two component elution case shown in Fig. 4.

Figure 5 shows the chromatogram for isocratic elution of a binary sample (component 1 and component 2) with a small ion (component 3) as the eluent (component 3) in inward flow RFC. The separation factors are constants in this case ($\alpha_{13} = 0.6$, $\alpha_{23} = 3.0$). This is similar to the binary elutions with a competing modifier in the mobile phase.

Multi-stage operations can also be simulated with our code. Figure 6 shows the effluent history of a reverse flow displacement process, in which the displacer (component 2) is introduced with a reversed flow direction after an incomplete period of frontal adsorption of component 1, which lasted $\tau = 3$. Such reverse flow displacement operation is actually very common in industrial AFC practice [32] aimed at improving process efficiency and reducing column clogging. It has been used in the elution stage of affinity chromatography by Chase [33]. In Fig. 6, the combination of outward flow adsorption and inward flow displacement gives better results, since its process time is slightly shorter.

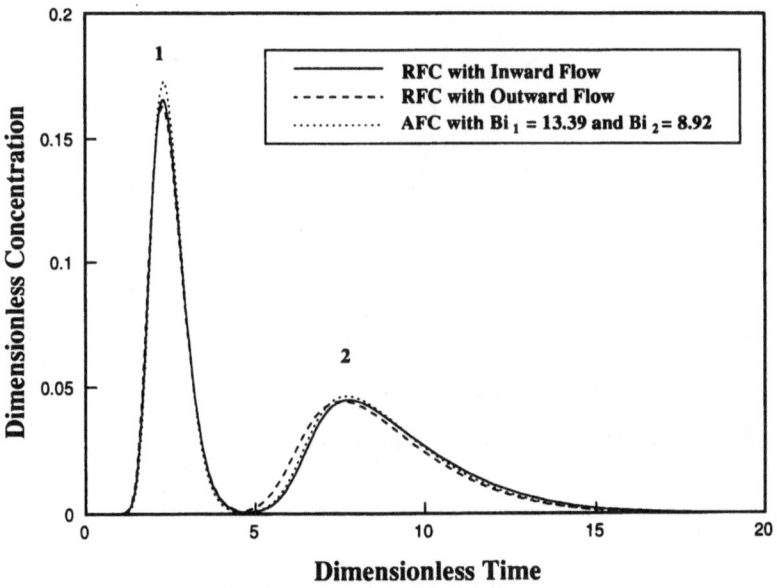

Fig. 4. Comparison of inward and outward flow RFC and AFC in elution

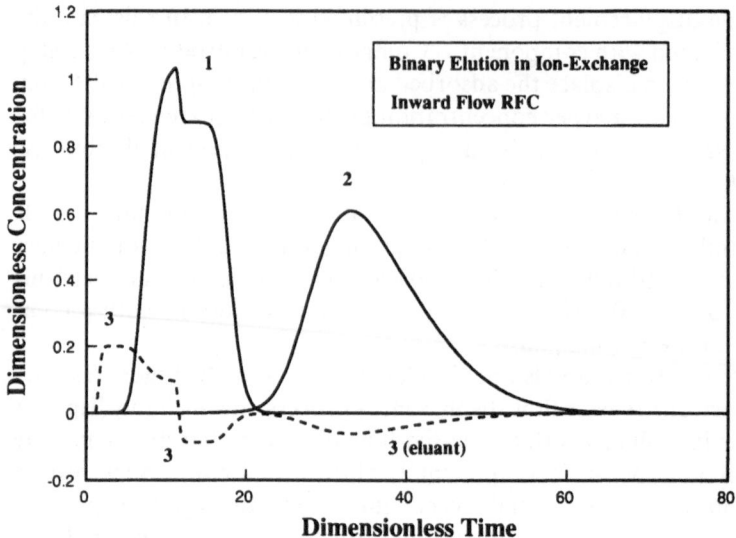

Fig. 5. Binary elution in ion-exchange RFC

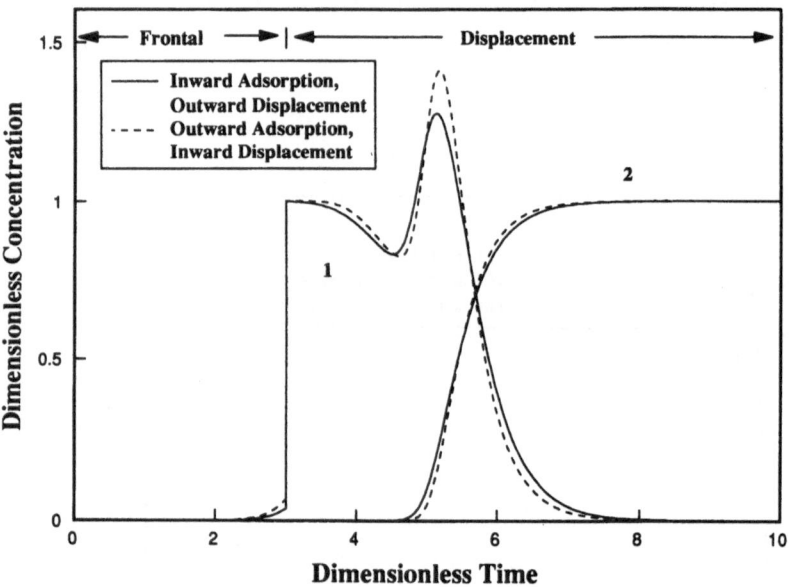

Fig. 6. Effect of RFC flow direction in reverse flow displacement

4.2 Effect of V_0

V_0 represents the ratio of the central cavity volume of the RFC column to the bed volume. The effect of its value on elution process is shown in Fig. 7, in which

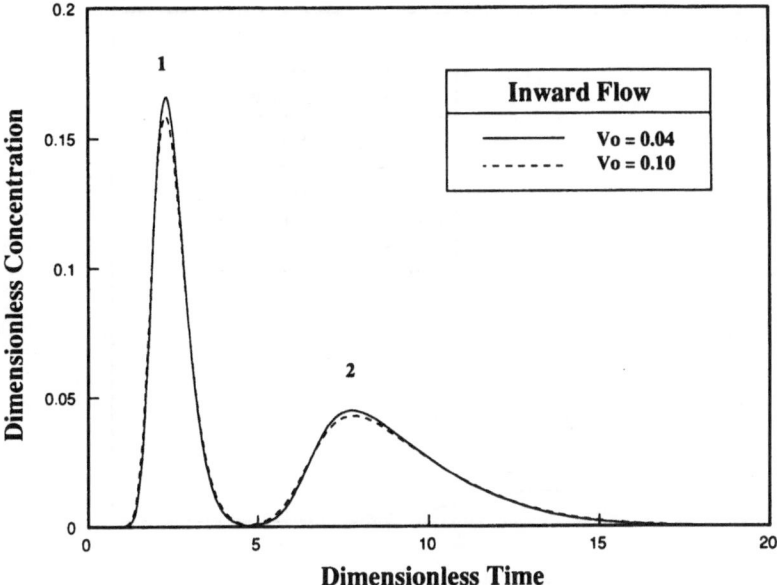

Fig. 7. Effect of V_0 on elution in inward flow RFC

the RFC column's X_1 and h are fixed while changing the X_0 value. In Fig. 7, the solid lines are the same as those in Fig. 4. The way V_0 values are changed from 0.04 to 0.1 requires that the Pe_i values for $V_0 = 0.1$ be reduced to 86.88% of those for $V_0 = 0.04$ case, and for η_i values the percentage is 94.54%. These two percentage values can easily be obtained by checking the changes in $(X_1 - X_0)$ and V_b and their relationship with Pe_i and η_i. Figure 7 shows that the peak heights are reduced and so is the peak resolution when V_0 is increased from 0.04 to 0.1. This has the same effect as reducing column length in AFC.

4.3 Effects of Pe_i, η_i and Bi_i on Elution

Figures 8 to 10 clearly show that the increase of Pe_i, η_i, or Bi_i value increases the peak heights and the peak resolution. In these three figures the solid line curves are the same as those in Fig. 4. In Fig. 8 $Pe_1 = Pe_2 = \infty$ case is plotted in order to show the errors that one may encounter if radial dispersion is neglected. In the case shown in Fig. 8 such errors are quite large. Figure 9 also indicates that the peaks are sharper if film mass transfer coefficients for the two components are larger since the dimensionless parameter η_i is proportional to k_i. Figure 11 shows that the resolution of the elution peaks is decreased if the intraparticle diffusion coefficients are reduced as reflected in the dimensionless parameters, η_i and Bi_i.

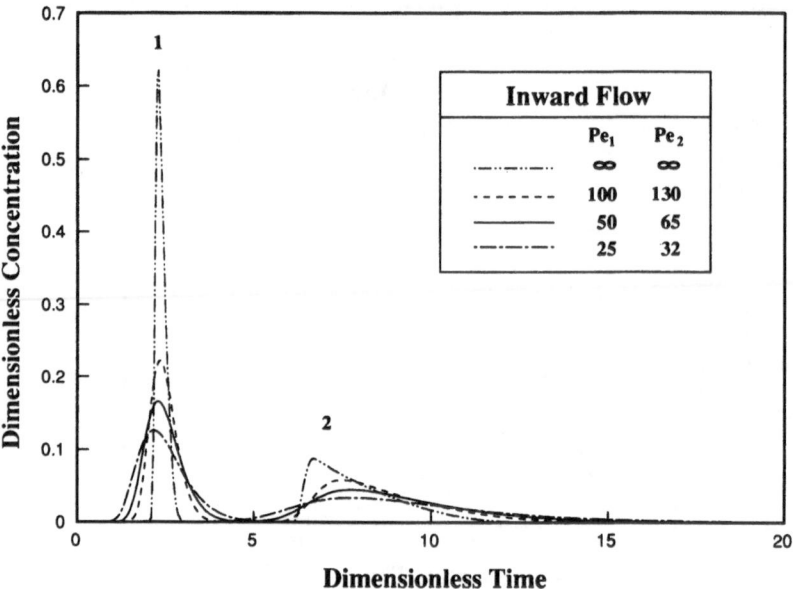

Fig. 8. Effect of Pe_i on elution in inward flow RFC

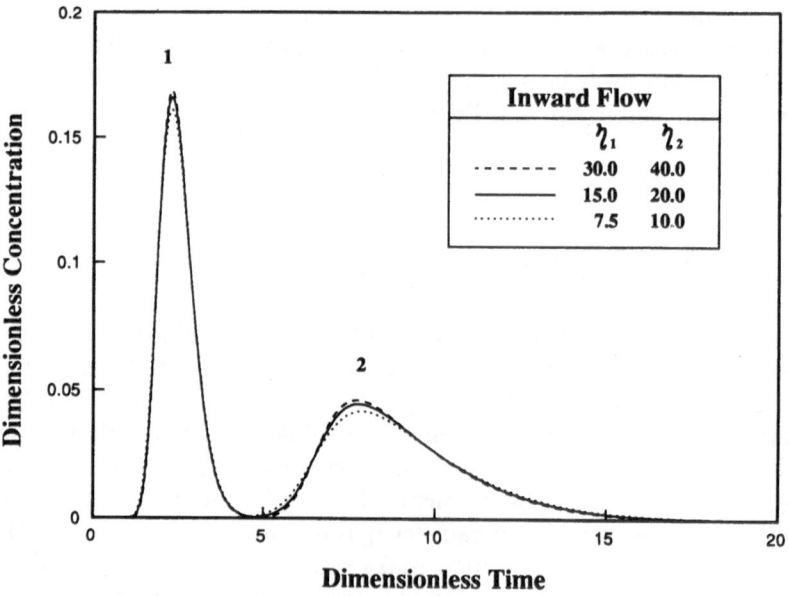

Fig. 9. Effect of η_i on elution in inward flow RFC

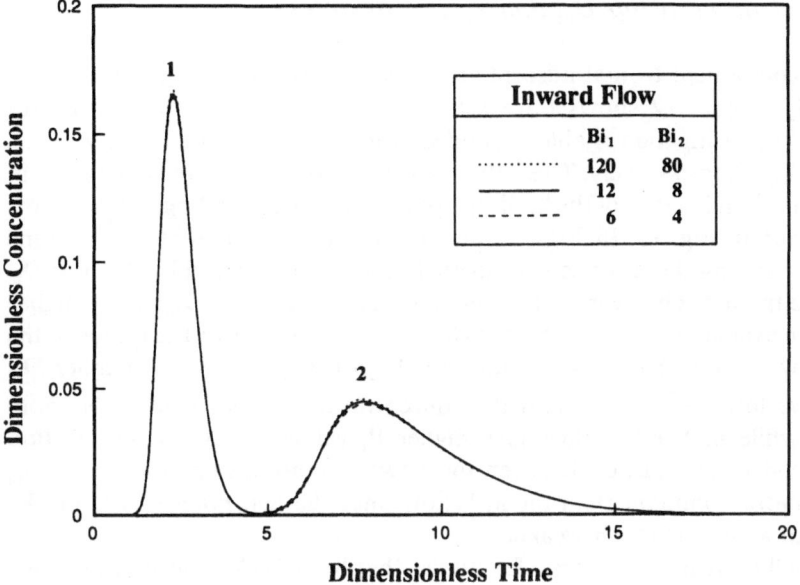

Fig. 10. Effect of Bi_i on elution in inward flow RFC

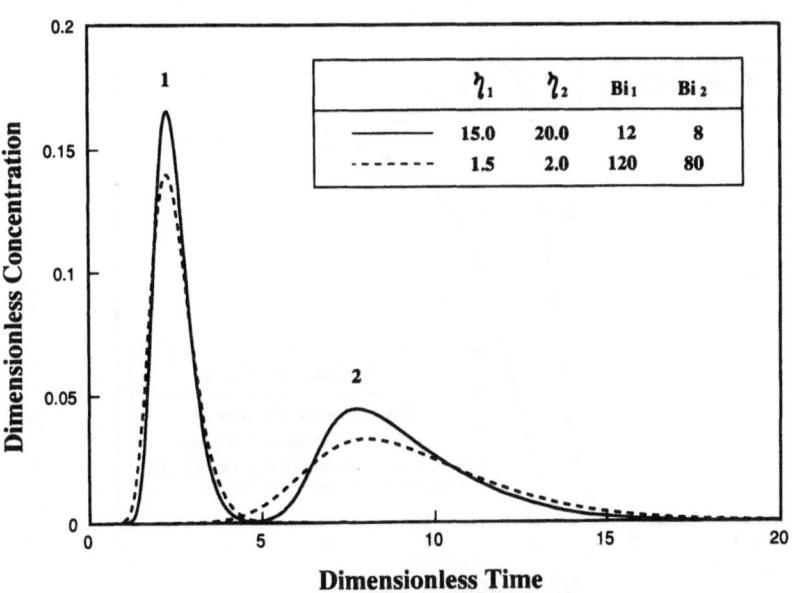

Fig. 11. Effect of D_{pi} on elution in inward flow RFC

4.4 Effect of Treating D_{bi} and k_i as Variables

The two component frontal adsorption system shown in Fig. 2 is chosen as a case study. Figure 12 shows three sets of inward flow breakthrough curves obtained from using the variable D_{bi} and k_i average values evaluated at $V = 0.5$, or $(X_1 + X_0)/2$, respectively. These three sets of curves show some differences in the sharpness and height of the "roll-up" peak. The corresponding outward flow case is given in Fig. 13. In Figs. 12 and 13, it is obvious that there are some differences among the three sets of curves in each figure caused by the way D_{bi} and k_i are treated. The average D_{bi} and ki evaluated at $(X_1 + X_0)/2$ are higher than those evaluated at $V = 0.5$ since $(X_1 + X_0)/2$ is closer to the center of the radial flow column than $V = 0.5$, and v is higher at $(X_1 + X_0)/2$. Higher D_{bi} values give lower $\overline{Pe}_i$ values and the concentration profiles tend to be more diffused, while higher k_i values give higher B_i values and the concentration profiles tend to be sharper. Between these two compromising factors, the D_{bi} factor is more dominant than the k_i factor, since the dependence of k_i on the coordinate X or V is much weaker.

It is well known in mass transfer studies that both Peclet and Biot numbers show some asymptotic behavior. When Pe_i values become larger, the system becomes less sensitive to the changes in Pe_i values. When Bi_i values are in the range of above 1, the higher the Bi_i values the less sensitive the system to the increase of Bi_i values. When Bi_i values are sufficiently small, the internal concentration gradient inside a particle can be neglected. On the other hand, when the Bi_i values are sufficiently large, the external interface mass transfer

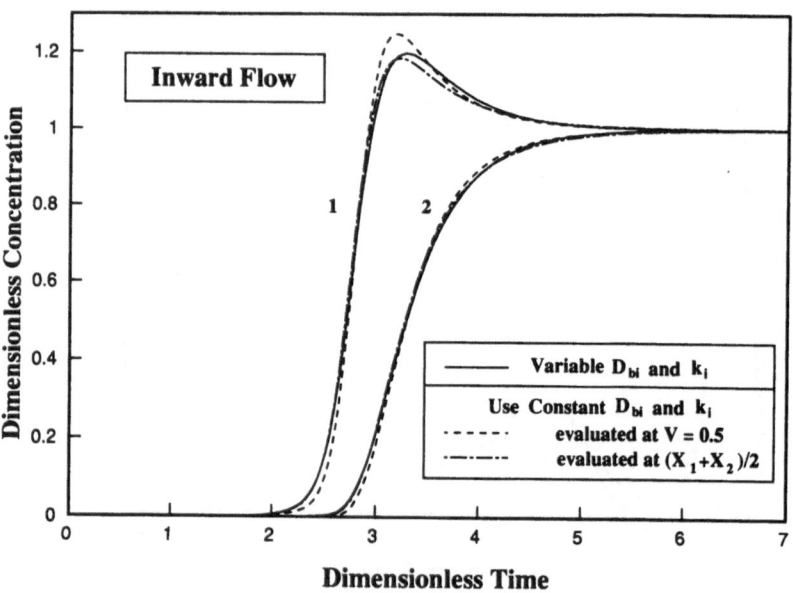

Fig. 12. Effect of treating D_{bi} and k_i as variables in inward flow RFC

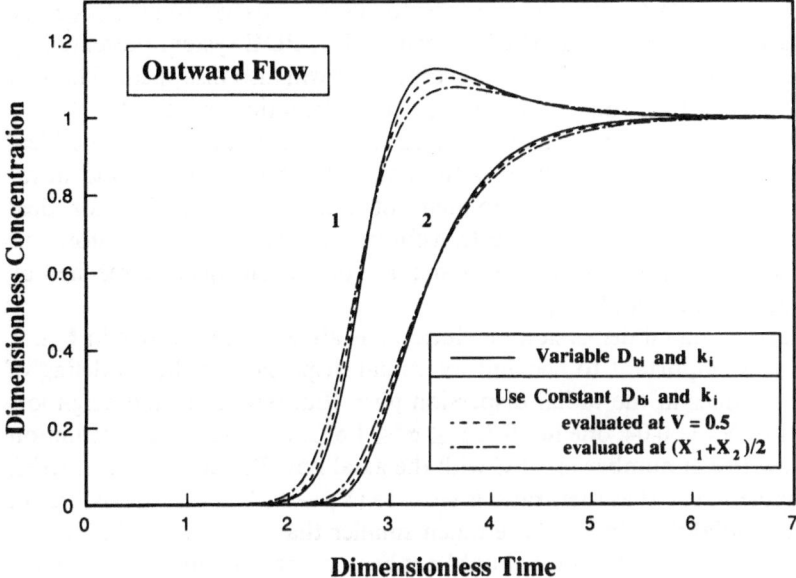

Fig. 13. Effect of treating D_{bi} and k_i as variables in outward flow RFC

resistances become negligible. These arguments are very helpful in determining the parameter ranges in which the treatment of D_{bi} and k_i as variables become important. Generally speaking, when the Pe_i values are large, the errors caused by using averaged D_{bi} values are small. If Bi_i values are not close to 1, then the treatment of k_i values as variables will have little effect. In most cases such averaging treatment causes some error, but they are not terribly severe. If an averaging treatment is necessary, such as in some analytical treatments in which simplification may be essential, one should not be inhibited from doing so. A good averaging method obviously can reduce error. It is difficult to provide exact general rules for averaging D_{bi} and k_i values, since the system is quite complex. From Figs. 12 and 13, and other extensive simulations, including different RFC operation and different flow directions, it is found that the averaging point should be at an X position farther away from X_0 than $(X_1 + X_0)/2$. $V = 0.5$ sometimes proves to be too close to X_1 as an averaging point, while at other times it becomes too far away, but generally speaking it is a better choice than $(X_1 + X_0)/2$. For numerical solutions it is desirable to treat D_{bi} and k_i as variable, since it is not only possible but also quite convenient if a suitable numerical procedure, such as the one presented in this chapter, is used.

4.5 Comparison of RFC and AFC

If the radial dispersion term in Eq. (9) is neglected, and ki values are treated as constants independent of the variation of v, i.e. ξ_i and Bi_i are treated as constants, then the RFC's dimensionless PDE system, Eqs. (9) and (10), become the same as the corresponding AFC's dimensionless PDE system presented by Gu et al. [28]. Inward and outward flow difference will also disappear. This can be easily verified by using a coordinate transformation with $V^* = 1 - V$. This conclusion is, of course, valid for more simplified cases, such as the ideal RFC which neglects radial dispersion, external film mass transfer, and intra-particle diffusion. Rhee et al. [24] pointed out that system equations for ideal radial flow chromatography can be transformed to the system equations for ideal AFC with Langmuir isotherms. A similar conclusion was also reached by Rice [34] and Huang et al. [7].

The effects of radial dispersion on elution has already been shown in Fig. 9. It is often very important to account for radial dispersion in the modeling of RFC. The α values in the radial dispersion term (Eq. (9)) are in the neighbor-hood of 1. For $V_0 = 0.04$, one has $0.328 \leq \alpha \leq 1.672$. Comparing the definition of radial flow Peclet number in RFC with the axial flow Peclet number in AFC, it is reasonable to say that their ratio is close to $(X_1 - X_0)/L$. Obviously, Peclet numbers for radial flow in RFC are much smaller than those for axial flow in AFC. A ratio between $1:20$ to $1:5$ should not be uncommon. Furthermore, RFC is designed primarily for preparative and production scale separations; thus Peclet numbers for radial flow tend to be even smaller. Typical values may often be below or not very far above 100. This is in agreement with the estimation by Tharakan and Chau [17] in their study of a radial flow bio-reactor for mammalian cell culture. In preparative and large scale AFC, Peclet numbers often reach hundreds or higher and neglecting axial dispersion (i.e. assuming Peclet numbers equal to infinity) often does not give large errors. For RFC, one may not be able to make such assumptions without risking substantial errors. Thus, one should be very cautious when assuming negligible radial dispersion in RFC. In the discussion above it has been mentioned that, in RFC, radial dispersion coefficients are inversely proportional to the radial coordinate X. This is actually a very important identity of RFC arising from its special flow geometry, which also differentiates RFC from AFC in terms of dimensionless mathematical expressions. Neglecting radial dispersion in RFC means that its identity in mathematical modeling is partially lost.

As already shown in Figs. 3 to 5, if one sets the corresponding dimensionless parameters and isotherm expressions the same for both RFC and AFC, their differences in simulated effluent histories are similar. In order to have similar dimensionless constants for an AFC and an RFC, the AFC column should be a short one. Note that the dimensionless parameter V_0 is unique in RFC, and for the corresponding AFC one needs to pick the Bi_i values from the variable Bi_i values of RFC. Because of this kind of close analogy, some data obtained from a short AFC column may be used for reference in RFC. This also helps in transforming an existing AFC setup to RFC. It is also obvious that many studies for AFC, such as multicomponent interference, can be qualitatively applied to RFC.

The difference between RFC and AFC may arise from the differences in the dimensionless parameters, especially Pe_i and η_i. In reality, Pe_i and η_i values for

RFC columns are usually several times smaller than in a longer AFC column. Notice that η_i values are proportional to the "dead volume time" of the bed and this time value for RFC is often several times smaller than in AFC because of the shorter flow path in RFC. Figures 9 and 10 show that both Pe_i and η_i values are very important to the sharpness of the concentration profiles and peak resolutions. The higher the Pe_i and η_i values the better resolution. Thus, generally speaking, RFC provides lower resolution than does AFC. This is why RFC is not intended for analytical purposes. The effect arising from the difference of Bi_i values in RFC and AFC is not discussed here because of some uncertainties. Bi_i values for RFC can be either higher or lower than those in AFC depending on value ranges of v in RFC and AFC. RFC usually has a higher volumetric flow rate, but it does not necessarily have larger linear flow velocities since its cross-sectional flow area is much larger than that of AFC's.

The most important advantage of RFC over AFC with longer column is, again, that in RFC the cross-sectional area perpendicular to flow direction is very large and the flow path is relatively short. These two factors help reduce the pressure drop in the bed and permit a much higher flow rate, and thus promote productivity.

5 Extensions of the General RFC Model

All the extensions to the basic general multicomponent rate model for AFC by Gu [35] have also been applied to the RFC model. Such extensions include second order kinetics, size exclusion effect and reaction in the liquid phase for modeling of biospecific elution using soluble ligand. These extensions have been carried out with ease. Details are omitted here, since the necessary modifications for adding second order kinetics involving only the particle phase governing equation, in which the AFC and the RFC model do not differ except that the k_i values in RFC are variables. The addition of reaction terms for macro-molecule and soluble ligand interaction involves the bulk-fluid phase, but it does not touch the characteristic terms of the RFC model. It has also been easily implemented [35].

6 Summary

A general nonlinear multicomponent rate model for RFC has been developed. Radial dispersion and mass transfer coefficients are treated as variables in the model. The model is solved numerically by using finite element and orthogonal collocation methods for the discretizations of bulk-fluid and particle phase PDEs, respectively. Various chromatographic operations have been simulated.

The diffusional and mass transfer effects on the RFC elution process have been studied. It has been found that for simple chromatographic operations inward flow is generally better than outward flow in RFC, since inward flow generally provides sharper concentration profiles. The treatment of radial dispersion and mass transfer coefficients by using averaged values instead of treating them as variables causes some errors and such errors may be reduced by properly taking the averages.

In numerical calculations treating the radial dispersion and mass transfer coefficients as variables adds very little complexity if the finite element method is used for the discretization of the bulk-fluid phase governing equation. It has been found that, in contrast to AFC, dispersion in the flow direction is often very important in RFC. The dynamic RFC behavior is similar to that of AFC with a shorter column, which has low Pe_i and η_i in non-dimensional analysis values. The theoretical treatment of RFC with comparison to AFC provides some useful information and it is helpful for the scale-up of RFC, either from a smaller scale RFC or from AFC to RFC. The general model has also been extended to include second order kinetics, size exclusion effect and reaction in the liquid phase for modeling of biospecific elution using soluble ligand. The numerical procedure presented in this chapter also serves as an example of how to deal with fixed bed problems involving variable physical properties.

7 References

1. McCormick D (1988) Bio/Technology 6:158
2. Saxena V, Weil AE, Kawahata RT, McGregor WC, Chandler M (1987) Am Lab (Fairfield, CT) 19:112
3. Ernst P (1987) Aust J Biotechnol 2:22
4. Chen H-L, Hou KC (1985) In: Tsao GT (ed) Annual reports on fermentation process. Academic, New York
5. Saxena V, Weil AE (1987) Bio Chromatography 2:90
6. Huang HS, Roy S, Hou C, Tsao GT (1988) Biotechnol Prog 4:159
7. Huang HS, Lee W-C, Tsao GT (1988) Chem Eng J 38:179
8. Plaigin M, Lacoste-Bourgeacq JF, Mandaro R, Lemay C (1989) Colloque INSERM 175:169
9. Lee WC, Tsai G-J, Tsao GT (1990) ACS Symp Series 427:104
10. Liapis AI (1989) J Biotech 11:143
11. Yang X, Tsai G-J, Tsao GT (1988) Third Chemical Congress of North America and 195th ACS National Meeting. Toronto
12. Balakotaiah V, Luss D (1981) AIChE J 27:442
13. Strauss A, Budde K (1978) Chem Techn 30:73
14. Hlavacek V, Votruba J (1977) In: Lapidus L, Amundson NR (eds) Chemical reactor theory: A review. Prentice-Hall, Englewood Cliffs
15. Chang H-C, Saucier M, Calo JM (1983) AIChE J 29:1039
16. Lepez de Ramos AL, Pironti FF (1987) AIChE J 33:1747
17. Tharakan J, Chau PC (1987) Biotechnol Bioeng 29:657
18. Lapidus L, Amundson NR (1950) J Phys Colloid Chem 54:821
19. Rachinskii VV (1968) J Chromatogr 33:234
20. Inchin PA, Rachinskii VV (1977) Russ J Phys Chem 47:1331
21. Lee W-C, Huang SH, Tsao GT (1988) AIChE J 34:2083

22. Kalinichev AI, Zolotarev PP (1977) Russ J Phys Chem 51:871
23. Lee WC (1989) PhD Thesis, Purdue University, West Lafayette
24. Rhee H-K, Aris R, Amundson NR (1970) Philos Trans R Soc (London) Ser A 267:419
25. Helfferich F (1962) J. Am Chem Soc 84:3242
26. Slattery JC (1981) Momentum, energy and mass transfer in continua. McGraw-Hill, New York
27. Gu T, Tsai G-J, Tsao GT (1991) Chem Eng Sci 46:1279
28. Gu T, Tsai G-J, Tsao GT (1990) AIChE J 36:781
29. Finlayson BA (1980) Nonlinear analysis in chemical engineering. McGraw-Hill, New York
30. Reddy JN (1984) An introduction to the finite element method. McGraw Hill, New York
31. Weber SG, Carr PW (1989) In: Brown PR, Hartwick RA (eds) High performance liquid chromatography. Wiley, New York
32. Ruthven DM (1984) Principles of adsorption and adsorption processes. Wiley, New York
33. Chase HA (1985) In: Vedrrall MS (ed) Discovery and isolation of microbial products. Ellis Horwood, Chichester
34. Rice RG (1982) Chem Eng Sci 37:83
35. Gu T (1990) PhD Thesis, Purdue University, West Lafayette

Artificial Neural Network Modeling of Adsorptive Separation

Mei-Jywan Syu[1], Gow-Jen Tsai[3], and George T. Tsao[2]*
[1] Department of Chemical Engineering and
[2] Laboratory of Renewable Resources Engineering
Purdue University, 1295 Potter Center, West Lafayette, IN 47907-1295, USA
[3] American Cyanamid Company, Medical Research Division, Pearl River, NY 10965, USA

Neural network computation, a new methodology, the concept of which originated earlier this century and is also known as parallel distributed processing (PDP), has been investigated by people from a wide range of backgrounds. Most applications of neural networks are located in various aspects of human cognition, including perception, memory language, and higher-level thought processes. Chemical engineers have paid attention to the neural network approach. This method has been implemented in the complex chemical systems. One of these chemical reaction systems was the separation of multiple components by adsorption where different adsorption kinetics, either theoretical or empirical, had been proposed and attempts to solve the complicated mechanisms of these systems had been made though few of them had obtained satisfactory results mostly due to incomplete or incorrect consideration of interaction between different compounds. The major advantages of most neural network approaches are the learning ability of non linearity, the tolerance to unpredicted noise (or experimental errors) through distributed memory, and the capability of retrieving appropriate information from the memory. Regarding the drawback inherent in the theory-based modeling of the multiple component adsorption and the purported capability of the neural network method to identify a system, neural networks were applied to these multicomponent

---

* To whom correspondence should be addressed

Advances in Biochemical Engineering
Biotechnology, Vol. 49
Managing Editor A. Fiechter
© Springer-Verlag Berlin Heidelberg 1993

adsorption systems to test their performance in the simulation of such nonlinear systems. This short article is a brief review of the very first application of this new methodology to the modeling of seven multicomponent adsorption systems – butyric acid, acetic acid, butanol (single-component), butyric acid/acetic acid, acetic acid/butanol, butyric acid/butanol (two-component), and butyric acid/acetic acid/butanol (three component). Polyvinylpyridine (PVP) resin was used as the adsorbent for these adsorption systems. Back-propagation neural network models (BPNN) with a saturation-type transfer function, instead of the common sigmoid transfer function, were successfully used in this study. The results show that the isotherms obtained from the neural network approach correlate with the experimental data significantly better than the multicomponent Langmuir isotherms. The interference between compounds has also been identified. The Langmuir equation for multi-component adsorption considering only competition among components is not adequate for describing the complicated behavior of such systems.

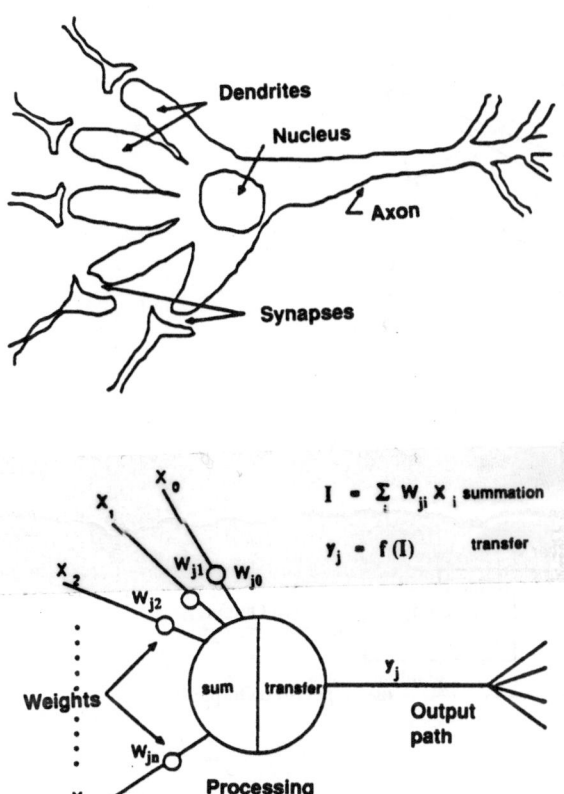

Fig. 2. The similarity between a human neuron and an artificial neuron

tungsten arc welding process. In the same year, Venkatasubramanian and Chan [25] applied the BPNN to fault diagnosis of a fluidized catalytic cracking unit.

There have been a number of theories developed to describe adsorption. For monolayer adsorption (chemisorption), the Langmuir model [26] is based on the assumption of homogeneous sites with each site for only one adsorbate molecule and no interaction between adsorbed molecules on neighboring sites. A model involving the ideal adsorption with weak interaction between adsorbed molecules on neighboring sites was proposed by Lacher [27] and by Fowler and Guggenheim [28]. The Langmuir model was modified with a power law expression of Freundlich [29]. Among various adsorption models, the Langmuir isotherm is still most widely applied. The Langmuir isotherm is valid, however, only if solutes are in simple competition for the adsorption sites involving no other interactions. It becomes invalid for solutes that are adsorbed onto a heterogeneous adsorbent with multi-adsorption sites. In such cases, none of the existing theories is sufficient for obtaining accurate multicomponent isotherms.

(a)

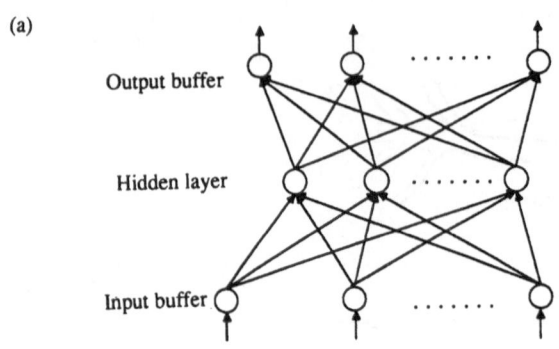

(b)

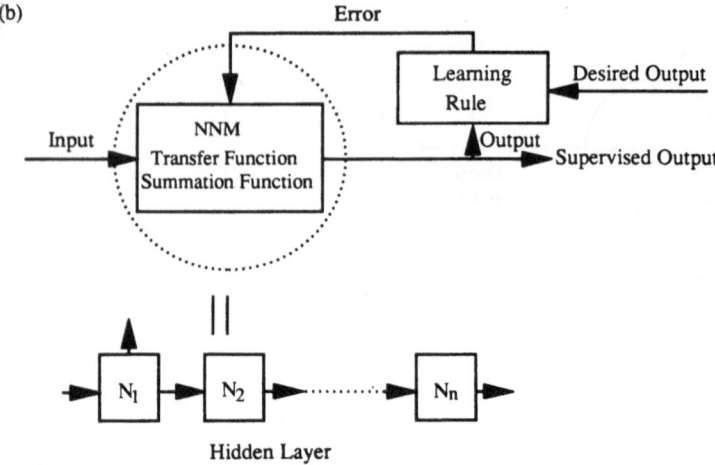

Fig. 3a. A BPNN structure **b** Basic computation of a BPNN

Hence, empirical correlations have been employed in multicomponent adsorption/desorption systems.

In this work, a BPNN was applied to the adsorption of butyric acid, acetic acid, and butanol onto a resin, polyvinylpyridine (PVP), which is a copolymer of 25 wt% of divinylbenzene and 75 wt% of vinylpyridine. Seven single and multiple component adsorption systems involving butyric acid, acetic acid, and butanol have been studied. Back-propagation models with a saturation-type transfer function were constructed to obtain the multicomponent adsorption isotherms. The comparison of the results from this saturation-type function was made with those from the Langmuir isotherms. The simulation and prediction results from BPNN models were demonstrated. The isotherms from BPNN

4 Results and Discussion

4.1 Simulated Back-Propagation Neural Networks for the Adsorption Systems

Back-propagation neural network models were designed and tested to get the optimal structures for these adsorption systems. The simulation from a single-component system to a three-component system will be discussed as follows.

4.1.1 Comparison of Transfer Functions

The sigmoid transfer function is always widely employed in back-propagation neural networks. In this study, neural network modeling was initially started with the sigmoid function. This function was tested and found to show weak simulation results for single-component adsorption systems. In Figs. 4(a), (b), and (c), the dashed lines, show a large simulation deviation near the origin. Modification of the network structures did not improve the situation to any significant extent. A saturation-type function was then used in the BPNN algorithm and structure to model the adsorption. The solid lines in Figs. 4(a), (b), and (c) represent the simulation of single-component adsorption by BPNN with this saturation transfer function. BPNN combined with this transfer function demonstrated a tremendous improvement over that with the sigmoid function.

4.1.2 Single-Component Adsorption

Neural network modeling was then done with this saturation-type transfer function (Fig. 5). The networks, with the average relative errors (a.r.e.) being 7.0% for the butyric acid system, 3.8% for the acetic acid system and 16.9% for the butanol system, simulate the behavior of these single-component adsorption systems very well. Even when some of the data points deviate significantly from expected values due to experimental errors, the neural networks are still able to track the correct behavior.

Three PEs in the hidden layer were used in modeling the single-component adsorption systems. When more than three PEs were employed in the hidden layer, more computation time was required to train the neural network but without better convergence. For the neural network with less than three PEs in the hidden layer, although it can converge, it cannot reach the convergence criterion as the one with three PEs. With PEs less than three, the mimimum state which this neural network can reach is not the optimal.

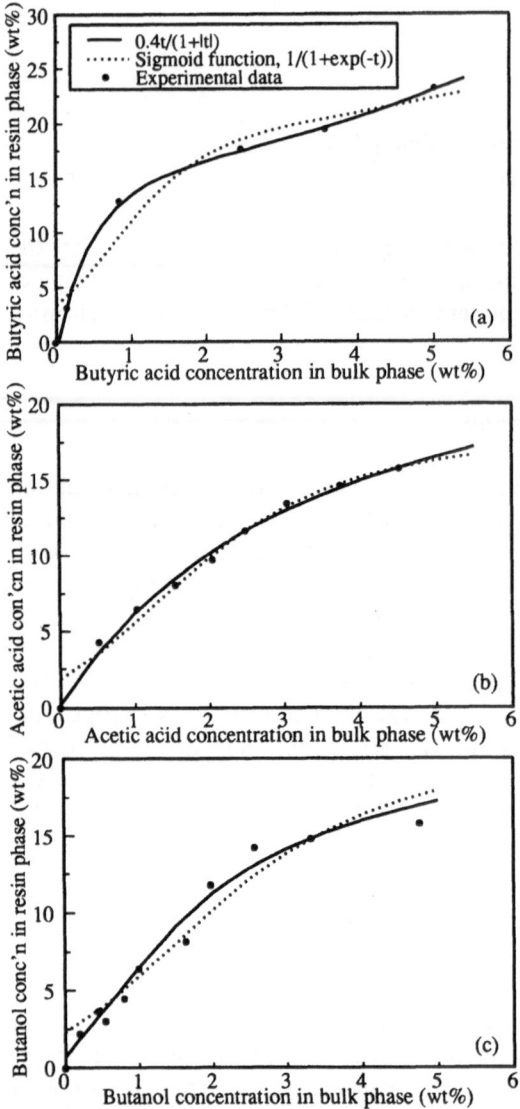

Fig. 4a–c. Comparison of saturation-type and sigmoid transfer functions in BPNN modeling of single-component adsorption systems. **a)** Butyric acid **b)** Acetic acid **c)** Butanol

4.1.3 Two-Component Adsorption

Figures 6(a), 8(a), and 10(a) show the simulation results of the two-component adsorption systems using experimental data for training constructed neural networks. The adsorption systems include binary solutions of butyric

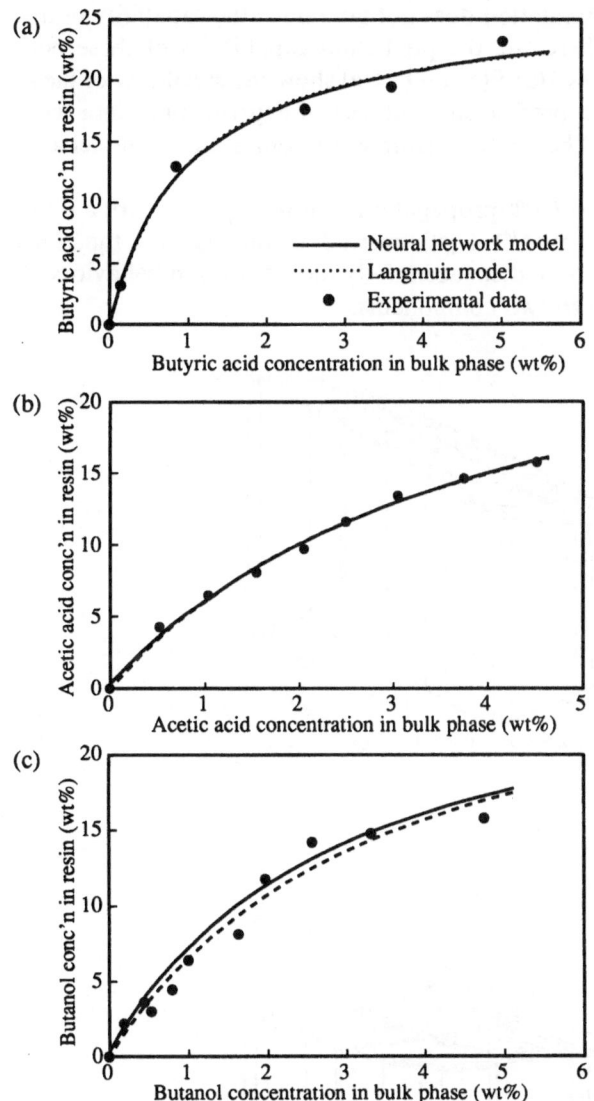

Fig. 5a–c. Comparison of neural network models and Langmuir isotherms for single-component adsorption systems. **a)** Butyric acid, **b)** Acetic acid, **c)** Butanol

acid/acetic acid, butyric acid/butanol, and acetic acid/butanol. The database for these systems was the combination of the experimental data from both single-component and two-component systems. For example, the data-base for the acetic acid/butanol system was from the experimental data of acetic acid, butanol (singl-component) and acetic acid/butanol (two-component) systems. The neural networks constructed were 2-3-2s (two inputs, three PEs in the hidden layer and two outputs) for all of these two-component systems.

The sucess of the above simulation does not guarantee the capability of the corresponding prediction. Therefore, the prediction capabilities of these networks were also tested. Figures 7(a), 9(a), and 11(a) show the capability of these simulated neural networks to predict different sets of experimental data with respect to the same systems. The relative errors were from both the model and the experimental deviation.

So far, the results show that back-propagation neural networks, with a saturation-type transfer function, $0.4 t/(1 + |t|)$, instead of the sigmoid function, can not only simulate, but also predict successfully the adsorption behaviors of multi-component systems up to two components.

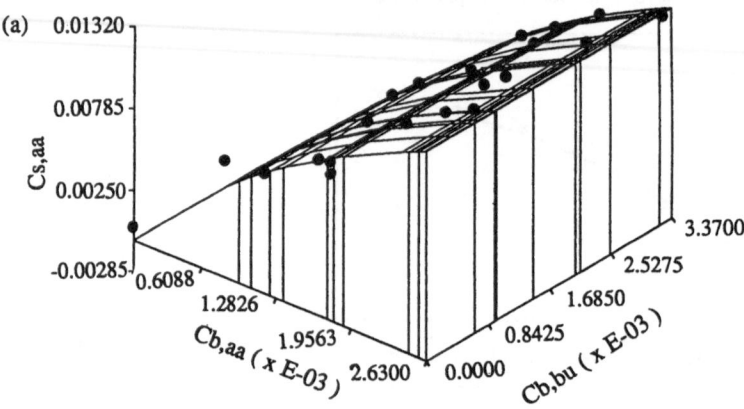

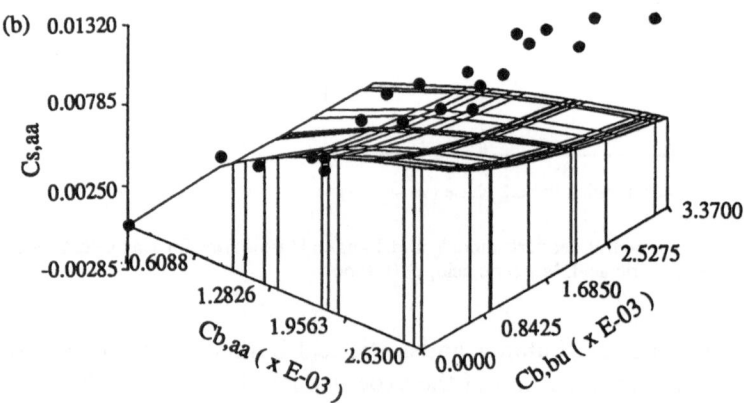

Fig. 6a, b. Comparison of the simulation results of acetic acid/butanol adsorption system from neural network and Langmuir isotherm. **a)** BPNN **b)** Langmuir model. The simulated surfaces are from neural network and Langmuir model. The *black circles* are the experimental points. Cs,i = concentration of compound i in resin phase, i = aa (acetic acid), bu (butanol). Cb,i = concentration of compound i in bulk phase, i = aa, bu

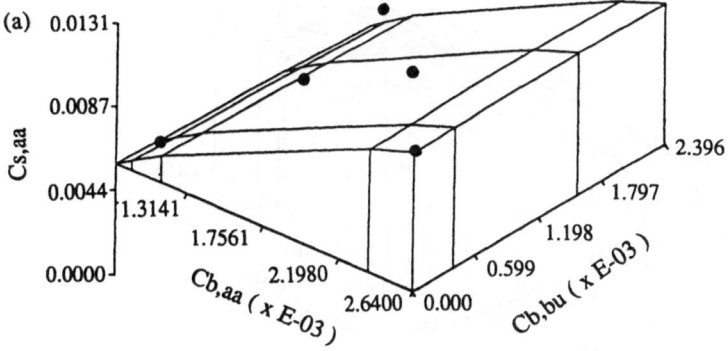

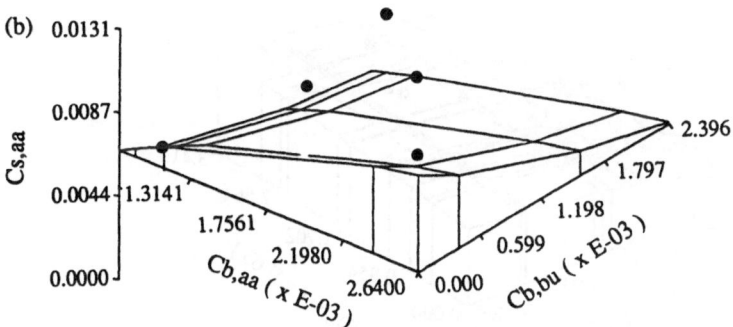

Fig. 7a, b. Comparison of the prediction results of acetic acid/butanol adsorption from neural network and Langmuir isotherm. **a)** BPNN **b)** Langmuir model

4.1.4 Three-Component Adsorption

Fifty-one experimental data points from seven different systems: butyric acid, acetic acid, butanol, butyric acid/acetic acid, butyric acid/butanol, acetic acid/butanol and butyric acid/acetic acid/butanol were used to train neural networks for the three-component system. Figures 12(a) and 12(b) are the simulation and prediction results of this three-component system. Average relative errors were used as the indicators for both the simulation and prediction results. Instead of giving a large set of simulation data, the results are represented by error distribution and plotted as bar charts in these figures.

Since the trained data file provided for the simulation and prediction of the three-component adsorption system, butyric acid/acetic acid/butanol, includes the data from the three-component system, as well as from single- and two-component systems, the success of modeling in this three-component system

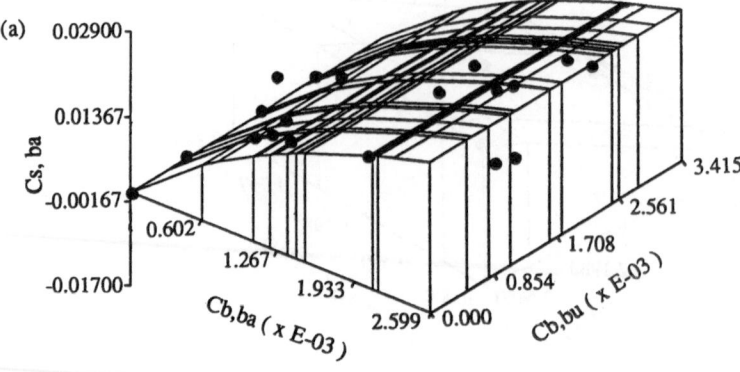

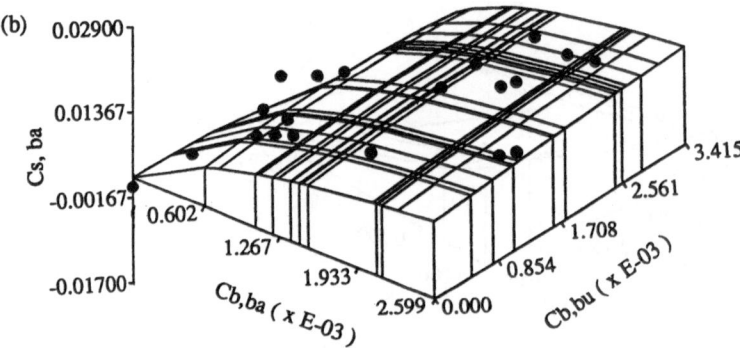

Fig. 8a, b. Comparison of the simulation results of butyric acid/acetic acid adsorption from neural network and Langmuir isotherm. **a)** BPNN **b)** Langmuir model

means that this simulated 3-2-3 neural network is able to model and predict this whole series of adsorption systems from a single-component to three-components. Therefore, the 1-3-1 BPNNs for single-component systems and the 2-3-2 BPNNs for those with two-components could be totally included in a model of 3-2-3 BPNN. Therefore, this 3-2-3 BPNN is sufficient to describe this whole group of adsorption systems well.

4.2 Langmuir Adsorption Isotherms

On the basis of the same experimental data files used for the neural network modeling, the adsorption equilibrium constants in Langmuir models were searched by an optimal subroutine from IMSL. The results are shown in Table 2. Equations, (10.1), (10.2), and (10.3), given in the table are the simulated

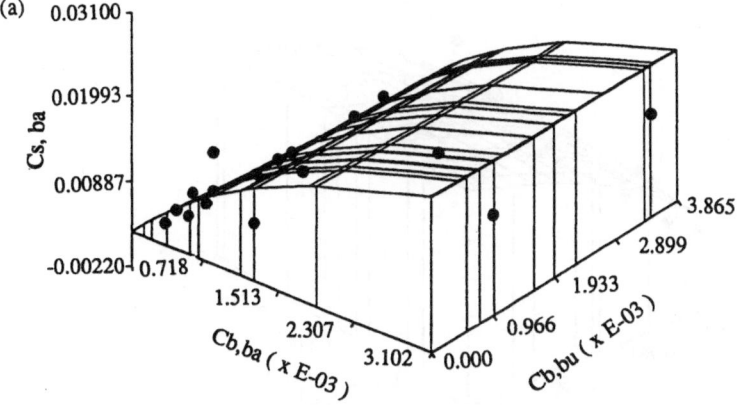

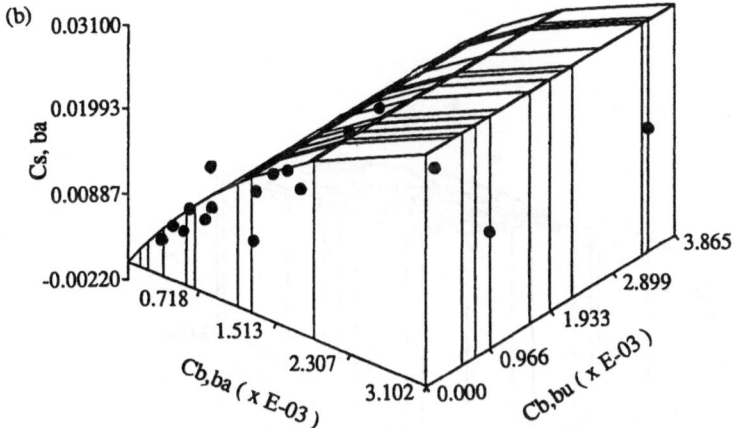

Fig. 9a, b. Coimparison of the prediction results of butyric acid/acetic acid adsorption from neural network and Langmuir isotherm. a) BPNN b) Langmuir model

Langmuir models for butyric acid, acetic acid and butanol single-component adsorption systems, respectively. The adsorption models, Eqs. (11.1), (11.2), (12.1), (12.2), (13.1), and (13.2) for two-component systems were also constructed based on the assumption of Langmuir isotherms (Table 3). Similarly, the Langmuir models, Eqs. (14.1), (14.2), and (14.3), for the butyric acid/acetic acid/butanol three-component adsorption system are illustrated in Table 4.

4.3 Comparison of BPNNs and Langmuir Adsorption Isotherms

In this work, the simulation and prediction of the multicomponent adsorption systems by neural networks have proven their success. However, how well the

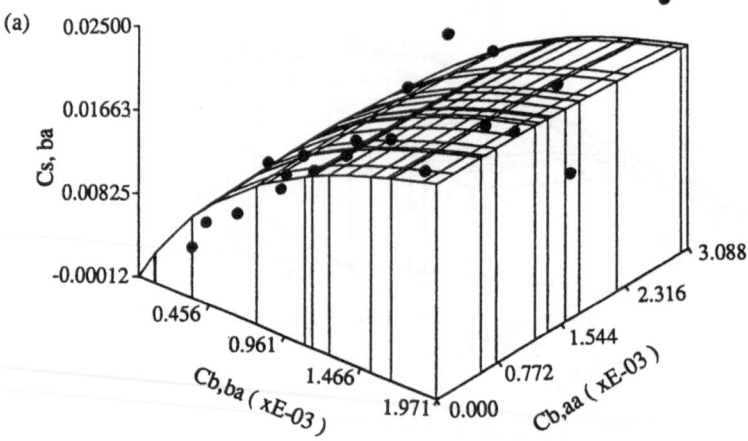

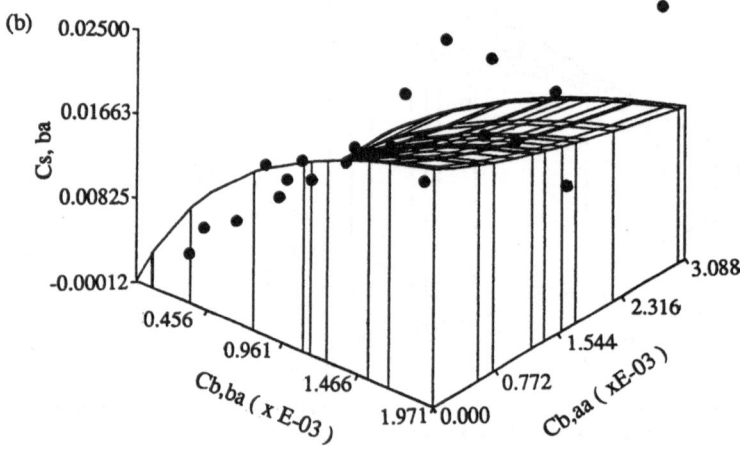

Fig. 10a, b. Comparison of the simulation results of butyric acid/butanol adsorption from neural network and Langmuir isotherm. **a)** BPNN **b)** Langmuir model

neural networks can simulate and predict these adsorption isotherms as compared to the multiple component Langmuir equation was examined further. The following simulation, using Langmuir isotherms, was carried out to obtain the results for comparison of both methodologies.

4.3.1 Single-Component Adsorption

The simulations for single-component systems were given in Figs. 5(a), (b), (c). The results from neural networks are as good as those from Langmuir models.

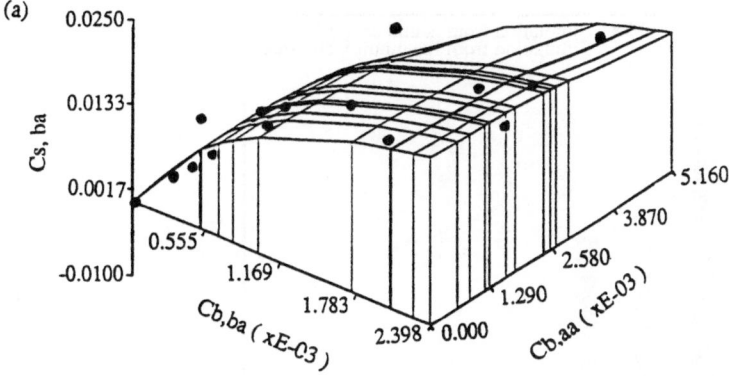

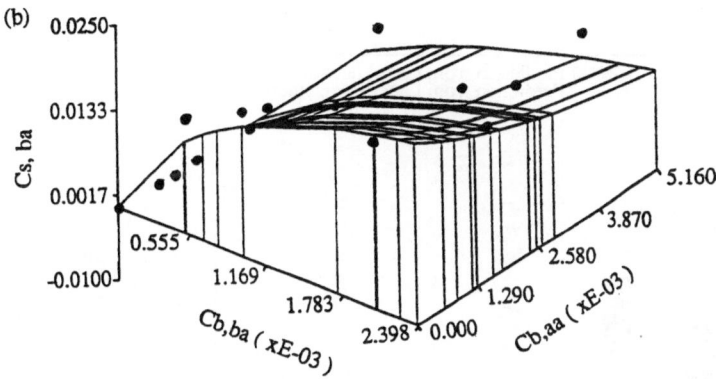

Fig. 11a, b. Comparison of the prediction results of butyric acid/butanol adsorption from neural network and Langmuir isotherm. a) BPNN b) Langmuir model

Actually, there is not much difference between these two approaches in modeling single-component systems. The success of the Langmuir models in simulating the adsorption behavior of these systems is no surprise because the Langmuir adsorption mechanism was originally postulated from single-component systems and known to work well for these systems. Meanwhile, the comparison in Figs. 5(a), (b), and (c) shows that neural networks can learn to track the true behavior and work equally well as Langmuir isotherms for the single-component systems in this study.

4.3.2 Multi-Component Adsorption

Figures 6–12 show the simulation and prediction results of the multicomponent systems from neural networks and Langmuir isotherms. As can be seen from the

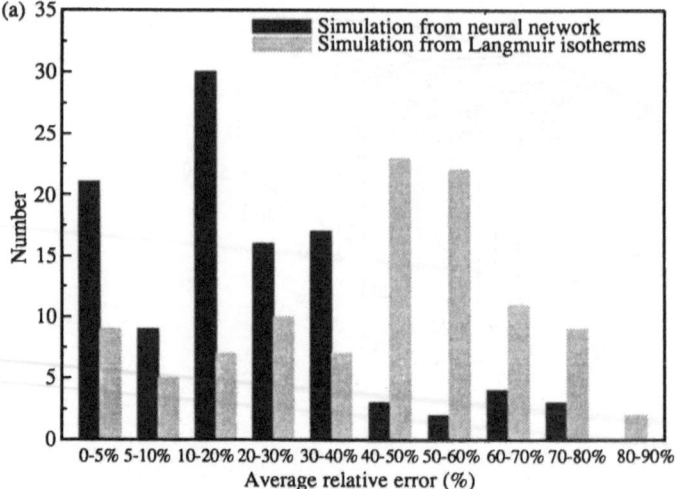

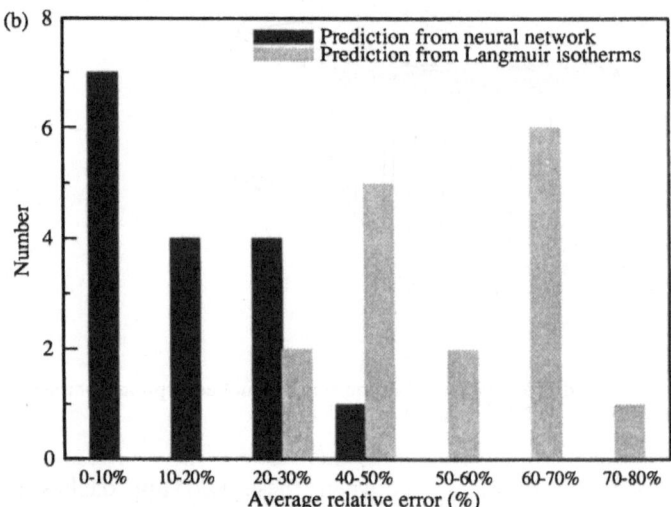

Fig. 12a, b. Comparison of the error distribution of BPNN and Langmuir modeling for butyric acid/acetic acid/butanol adsorption. **a)** Simulation **b)** prediction

figures, neural networks gave a more convincing performance than Langmuir models for these multicomponent systems. Neural networks always obtain less absolute relative errors than Langmuir models. A superior model can always simulate experimental facts but, in addition, predict more unknowns or the information that is lacking and difficult to obtain from experiments. In this way, many experiments could be avoided. Langmuir isotherms obviously cannot precisely predict the behavior of multicomponent adsorption because the theoretical background of this model includes only simple competition for active sites

Table 2. List of Langmuir equations for single-component adsorption systems with adsorption coefficients simulated from experimental data

Langmuir isotherms for single-component adsorption system

Butyric acid	$$C_{s,ba} = \frac{37.0933\,C_{b,ba}}{1 + 1064.92\,C_{b,ba}}$$	(10.1)
Acetic acid	$$C_{s,aa} = \frac{11.1854\,C_{b,aa}}{1 + 569.27\,C_{b,aa}}$$	(10.3)
Butanol	$$C_{s,bu} = \frac{8.05465\,C_{b,bu}}{1 + 758.324\,C_{b,bu}}$$	(10.3)

Table 3. List of Langmuir equations for two-component adsorption systems

Langmuir isotherms for two-component adsorption system

Acetic acid/Butanol	$$C_{s,aa} = \frac{11.1854\,C_{b,aa}}{1 + 569.27\,C_{b,aa} + 758.324\,C_{b,bu}}$$	(11.1)
	$$C_{s,bu} = \frac{8.05465\,C_{b,bu}}{1 + 569.27\,C_{b,aa} + 758.324\,C_{b,bu}}$$	(11.2)
Butyric acid/Acetic acid	$$C_{s,ba} = \frac{37.0933\,C_{b,ba}}{1 + 1064.92\,C_{b,ba} + 569.27\,C_{b,aa}}$$	(12.1)
	$$C_{s,aa} = \frac{11.1854\,C_{b,aa}}{1 + 1064.92\,C_{b,ba} + 569.27\,C_{b,aa}}$$	(12.2)
Butyric acid/Butanol	$$C_{s,ba} = \frac{37.0933\,C_{b,ba}}{1 + 1064.92\,C_{b,ba} + 758.324\,C_{b,bu}}$$	(13.1)
	$$C_{s,bu} = \frac{8.05465\,C_{b,bu}}{1 + 1064.92\,C_{b,ba} + 758.324\,C_{b,bu}}$$	(13.2)

Table 4. List of Langmuir equations for three-component adsorption system

Langmuir isotherms for three-component adsorption system

Butyric acid	$$C_{s,ba} = \frac{37.0933\,C_{b,ba}}{1 + 1064.92\,C_{b,ba} + 569.27\,C_{b,aa} + 758.324\,C_{b,bu}}$$	(14.1)
Acetic acid	$$C_{s,aa} = \frac{11.1854\,C_{b,ba}}{1 + 1064.92\,C_{b,ba} + 569.27\,C_{b,aa} + 758.324\,C_{b,bu}}$$	(14.2)
Butanol	$$C_{s,bu} = \frac{8.05465\,C_{b,bu}}{1 + 1064.92\,C_{b,ba} + 569.27\,C_{b,aa} + 758.324\,C_{b,bu}}$$	(14.3)

but no other interaction among compounds. Meanwhile, the comaprison in Figs. 6–12 suggests that neural networks are also superior in predicting these multicomponent adsorption systems.

4.4 Study of Neural Networks

The functions of the neural networks were examined. It was shown that the less the convergence criterion, the greater the number of iterations needed to reach the criterion (Fig. 13). When the convergence criterion is decreased to a certain value, the number of iterations increases drastically because it is getting closer to the minimum state during the training path. Increasing the number of iterations usually only makes the convergence state fluctuate around the minimum without producing further significant improvement.

4.4.1 Processing Element

An increase in the number of the processing elements (PEs) in the hidden layer also results in an increase in the number of iterations and computation time to reach the same convergence criterion (Fig. 14). However, if the number of PEs in the hidden layer reduces to a certain value, the neural network being constructed will not be able to describe the system well. On the other hand, if the number of PEs in the hidden layer increases to a certain value, the number of iterations will increase drastically without further reducing the average relative error. In such a case, the use of more PEs becomes redundant. There is only one optimal neural network structure for a specific problem, although there can be many neural networks constructed for a specific case.

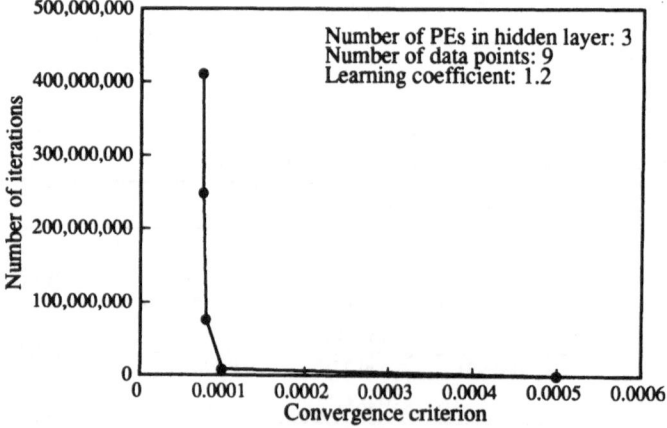

Fig. 13. Convergence criterion vs. number of iterations

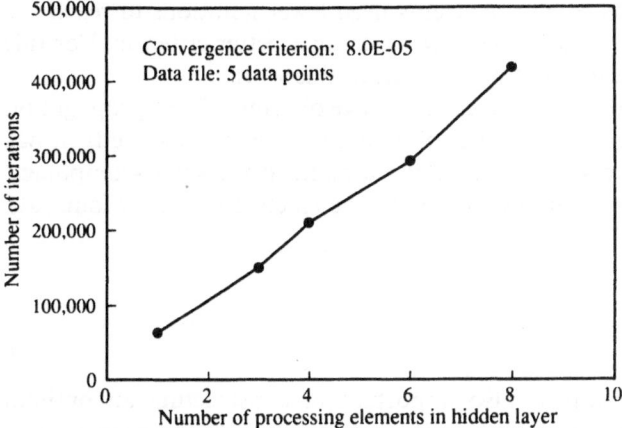

Fig. 14. Number of processing elements in hidden layer vs number of iterations

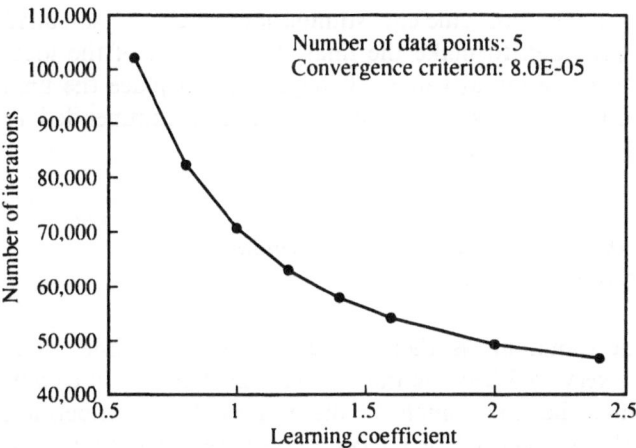

Fig. 15. Learning coefficient vs. number of iterations

4.4.2 Learning Coefficient

As the learning coefficient is increased, the number of iterations to reach the same convergence decreases (Fig. 15). The learning rate determines the step size in adjusting the connection weights. For a smaller learning rate, the step in each adjustment is smaller that results in a longer computation time to reach the desired minimum state. For a larger learning rate, it takes less time to reach the desired convergence. However, a large learning rate may easily result in the divergence of the solution if the system being studied does not have good stability. Though larger learning coefficients need fewer iterations to reach the same convergence, this may not guarantee reaching a better criterion. For this study, a learning coefficient of 1.2 was chosen in this work.

It has often been claimed that the disadvantage of using a back-propagation model is that it requires a large number of data inputs. From this study, it has not been shown to be necessarily true. For example, in the single-component adsorption, the same convergence criterion can be reached with only four data points.

4.4.3 Initial Guess

The initial guess for weights is also important in constructing an optimal network. A proper initial guess will help the network reach convergence while an improper initial guess might easily reach a worse local minimum. Especially for BPNN, the initial guess is a very important and sensitive decisive factor for determining an optimal structure. As the dimension of a network expands, the number of local minimum states increases. Therefore, it is more likely to reach an improper minimum unless an efficient stimulation, such as so-called simulated annealing, is provided to make the final state get out of the local point. A slight difference in the initial values of weights might place the final states far apart. A sensitivity analysis of initial weights is recommended to efficiently obtain a better initial guess.

4.5 Effect Among Compounds in Multicomponent Adsorption Systems

It has been proved that Langmuir isotherms are not able to model these multicomponent systems very well because this model considers only the competitive effect. It is assumed that there might be other effects than competition between compounds in these systems. The concentrations were first calculated based on totally no effect between the components. Then, the simulated concentrations are compared to the concentrations directly from the experiments. Thus, the effect of the adsorption capacity of each component caused by the existence of the other compound(s) in these real systems can be analyzed. For example, if the simulation results of the butanol concentrations, assuming no effect, are even smaller than the real concentrations in the multicomponent systems, then the adsorption capacity of butanol in the real butyric acid/butanol, acetic acid/butanol and butyric acid/acetic acid/butanol systems is enhanced by the existence of the other compound(s).

Table 5 summarizes the interference of these compounds in different multicomponent systems. It is clear from this table that the adsorption mechanisms in these multicomponent adsorption systems are too complicated to be solely described by the competition effect.

Table 5. The interactions between the compounds in two-component (butyric acid/acetic acid, butyric acid/butanol and acetic acid/butanol) and three-component adsorption systems (butyric acid/acetic acid/butanol)

	Butyric acid	Acetic acid	Butanol
Butyric acid	–	(m, c)	(c, e)
Acetic acid	(c, m)	–	(m, e)
Butanol	(e, c)	(e, m)	–

For three-component system, (butyric acid, acetic acid, butanol) = (m, c, e)
c: competition effect
e: enhancement effect
m: combination of both

5 Conclusions

In essence, this study shows for the first time the superiority of neural networks in modeling such complicated multicomponent adsorption systems. The back-propagation neural network was coupled with a saturation-type transfer function in this work. By this combination, the requirement for large data files could be avoided. Moreover, the results from this saturation-type function were significantly better than those from the sigmoid function. It provides a more efficient method than Langmuir isotherms in the modeling of the nonlinear adsorption. Even when there are experimental errors, neural networks are still able to recognize the errors and learn to adapt the data provided. In addition, a neural network is also capable of predicting the systems from the constructed model. The better prediction of a system is far more important than the success of simulation. In this study, neural networks have shown its robustness in modeling these adsorption systems. The interference among compounds in these adsorption systems beyond simple competition has also been clarified. A neural network is flexible not only in its structure, but also in the dimensions of both inputs and outputs. Giving the data files, neural networks can be constructed from different designs of input/output pairs according to different demands from the files provided. Besides, when using a traditional approach, several models are required to simulate the whole series of adsorption systems while only one single back-propagation neural network is sufficient to describe this whole series of adsorption systems, showing the high versatility of neural network modeling. This work represents, as far as we know, the first attempt at applying neural networks in modeling multiple component adsorption. The success of this attempt may stimulate additional interest in this new methodology to identify other separation as well as biosynthetic processes.

Acknowledgments. This work was supported by NSF Grants 8912150-BCS and EET 8712867. We would also like to acknowledge Fei Tan's experimental assistance.

6 References

1. Colvin G (1989) The C Users Journal 7:59
2. Neural Networks Professional II (1988) Neural Ware Inc., p 7
3. Hebb DO (1949) The organization of behavior. Wiley, New York
4. McCulloch WS, Pitts W (1943) Bulletin Math. Biophys. 5:115
5. Rosenblatt F Mechanization of Thought Processes: Proceedings of a Symposium held at the National Physical Laboratory, November 1958, London
6. Minsky M (1954) Neural nets and the brain-model problem. Thesis, Princeton University
7. Widrow G, Hoff ME (1960) Western Electronic Show and Convention, Convention Record
8. Barto AG, Sutton RS (1981) Biological Cybernetics 42:1
9. Hopfield JJ, Feinstein DI, Palmer RG (1983) Nature 304:158
10. Hinton GE, Sejnowksi TJ (1983) Proceedings of the Fifth Annual Conference of the Cognitive Science Society
11. Rumelhard DE, Zipser D (1985) Cognitive Science 9:75
12. Parker DB (1985) Technical Report TR-47, Institute of Technology, Center for Computational Research in Economics and Management Science, Cambridge MA
13. Minsky M, Papert (1969) Perceptrons, MIT Press, Cambridge MA
14. Anderson JA (1983) IEEE Transactions on Systems, Man, and Cybernetics 13:799
15. Marslen-Wilson WD, Welsh A (1978) Cognitive Psychology 10:29
16. Ballard DH, Hinton GE, Sejnowski TJ (1983) Nature 306:21
17. Cohen MA, Grossberg S (1983) IEEE Transactions on Systems, Man, and Cybernetics 13:815
18. Kohonen T (1982) In: Lang M (ed) Proceedings of the Sixth International Conference on Pattern Recognition, IEEE Computer Society Press, Silver Springs MD
19. Hopfield JJ, Tank D (1985) Biological Cybernetics 52:141
20. Tsirukis AG, Reklaitis GV, and Tenorio MF (1990) TR-EE 89–69, School of Electrical Engineering,
21. Werbos PJ (1988) Neural Networks 1:339
22. Cheok KC, Smith JC, Fernando JP (1989) Control and Computers 17:32
23. Lapedes A, Farber R (1987) Los Alamos National Laboratory Report LA-UR-87-2662
24. Karsai G, Andersen K, Cook GE, Ramaswamy K (1989) IEEE International Symposium on Intelligent Control. p 280
25. Venkatasubramanian V, Chan K (1989) AIChE J. 35:1993
26. Langmuir I (1916) J. Am. Chem. Eng. 38:2221
27. Lacher JR (1937) Proc. Roy. Soc. A161:525
28. Fowler RH Guggenheim EA (1939) Statistical thermodynamics. Cambridge University Press, Cambridge
29. Sips R (1948) J. Chem. Phys. 16:490
30. Rumelhart DE, McClelland (eds) Foundations, Chap 8 Vol. I, MIT Press, 1986
31. Yang X (1988) An adsorption-coupled acetone-butanol fermentation by *Clostridium acetobutylicum*. MS thesis, Purdue University, West Lafayette

Plate Models in Chromatography: Analysis and Implications for Scale-Up

A. Velayudhan[1] and M.R. Ladisch[1,2]
[1] Laboratory of Renewable Resources Engineering
[2] Department of Agricultural Engineering, Purdue University, West Lafayette, IN 47907, USA

Detailed chromatographic rate theories from the literature can be used to determine the appropriate plate count for a plate model of linear chromatography so that the bandspreading generated by the detailed rate model is reproduced by the plate model. This process provides a link between the plate count and the physical parameters that cause bandspreading. Each sample component can be assigned an appropriate plate count, thus allowing the accurate simulation of multicomponent separations even for widely differing adsorbates. Analytical solutions are presented for the Craig distribution and the continuous plate model for both finite-pulse elution and frontal chromatography. The Craig model is widely considered unsuitable because it assumes discontinuous flow; it is shown that, for a suitably corrected plate count, the Craig model is as accurate as the continuous-flow plate theory (except for the case of an unretained solute). Direct calculation of effluent histories from these plate models show excellent agreement between themselves and with results from complex rate models available in the literature. Reasonable agreement is also found when the plate models are used a priori to predict experimental scale-up results.

Advances in Biochemical Engineering
Biotechnology, Vol. 49
Managing Editor: A. Fiechter
© Springer-Verlag Berlin Heidelberg 1993

List of Symbols

c	mobile phase concentration, M
D_x	effective micropore diffusivity, $m^2 s^{-1}$
D_y	effective macropore diffusivity, $m^2 s^{-1}$
D_z	effective axial dispersion coefficient, $m^2 s^{-1}$
$I_x(a, b)$	incomplete beta function, $= \dfrac{\int_0^x t^{a-1}(1-t)^{b-1}\, dt}{\int_0^1 t^{a-1}(1-t)^{b-1}\, dt}$
J	plate count in continuous-flow plate model
k_f	external film mass transfer coefficient, $m\, s^{-1}$
K_i^d	dimensionless distribution coefficient; subscript i indicates i^{th} component
K_i^*	overall distribution coefficient, $= \varepsilon_y + (1 - \varepsilon_y)K_i^d$
k'	retention factor, dimensionless
L	column length, m
M	input pulse width (discretized)
N	number of spatial segments in the Craig simulation
N_{plate}	plate count
p_i'	probability of ith species being in the mobile phase
$P(a, x)$	incomplete gamma function, $= \dfrac{\int_0^x t^{a-1}e^{-t}\, dt}{\int_0^\infty t^{a-1}e^{-t}\, dt}$
Pe_x	microparticle Peclet number, $\dfrac{u(2r_x)}{D_x}$, dimensionless
Pe_y	pellet Peclet number, $\dfrac{u(2r_y)}{D_y}$, dimensionless
Pe_z	bed Peclet number, uL/D_z, dimensionless
$\overline{Pe}$	equivalent Peclet number in the axial-dispersion model
q	stationary phase concentration, M
q_i'	probability of ith species being bound to the stationary phase
r_x	microparticle radius, m
r_y	pellet radius, m
Sh	Sherwood number, $\dfrac{k_f(2r_y)}{D_y}$, dimensionless
t	time, s
t_R	retention time (first moment), s
Δt	time increment in the Craig simulation, s
u	superficial velocity, $m\, s^{-1}$
v	mobile phase velocity, $m\, s^{-1}$
v_{int}	interstitial velocity, $m\, s^{-1}$
x	axial coordinate, m
Δx	axial space increment, m

Subscripts

F	frontal chromatography
i	i^{th} component
j	j^{th} plate
k	k^{th} time interval

Greek Symbols

α_i	coefficient in the continuous-flow plate solution, $= \dfrac{N_i}{t_{R,i}}$, s^{-1}
δ_x	scaled microparticle radius, $\dfrac{r_x}{L}$, dimensionless
δ_y	scale pellet radius, $\dfrac{r_y}{L}$, dimensionless
ε_x	micropore porosity, m^3 micropores per m^3 microparticles
ε_y	macropore porosity, m^3 macropores per m^3 pellet
ε_z	bed porosity, m^3 bed voids per m^3 bed
ϕ	volumetric phase ratio, dimensionless
μ_0	zeroth temporal moment, M s
μ_1	first temporal moment, s
μ_2	second temporal moment, s^2
$\bar{\mu}_2$	second central temporal moment, s^2
τ	input pulse width, s

1 Introduction

Analyses of linear chromatography are traditionally divided into two classes: the rate and the plate theories [1, 2]. The rate theories reflect the various mass-transfer and diffusion processes that occur in the chromatographic column in addition to sorption. The major advantage that these theories possess is their explicit dependence on the physical parameters involved in bandspreading, such as the film mass-transfer coefficient and the pore diffusivity. In contrast, the plate theories lump the various bandspreading contributions into a plate height that can conceptually be regarded as a measure of non-equilibrium. While this simplifies the mathematical treatment, the plate height is not directly related to the physical parameters that cause bandspreading, and these models are hence considered empirical [3].

It is possible to generate the equivalent of a plate height from the results of a rate model, and many such height-equivalent-to-a-theoretical-plate (HETP) expressions [1, 4–7] exist in the literature. Such HETP expressions have been used to examine how the bandspreading is affected by the column characteristics and operational variables, e.g., the van Deemter expression for HETP as a function of axial flow velocity [7].

General expressions for the effluent history are presented for both the Craig and the continuous-flow plate models for multicomponent linear frontal and elution chromatography. The drawback that is usually associated with the Craig model – that it substantially overestimates the resolution between peaks as well as the plate count – is shown to rest on an inappropriate definition of plate count. The Craig model can therefore be used wherever the continuous-flow plate model is applicable, except in the case of an unretained component, for which it erroneously predicts zero bandspreading. The relation between Craig plates and plug flow reactor segments is brought out. The analytical expressions for effluent history lend themselves to scale-up: the design parameters such as particle diameter and column dimensions can be varied, the resulting plate number calculated, and the appropriate plate solution (which is explicitly known) used to examine the quality of the separation. Since each step in this process has an explicit analytical expression, optimization is facilitated.

2 Theory

2.1 Theoretical Determination of Plate Count

Numerous solutions express H (the HETP) and the corresponding plate count N_{plate} (= L/H, where L is the column length) as the sum of contributions from various physical processes [3, 6, 8, 9]. Here we use Haynes' results [10]. The first moment (retention time) is given by

$$\mu_{1,i} = \frac{L}{u}[\varepsilon_z + (1 - \varepsilon_z)\varepsilon_y + (1 - \varepsilon_z)(1 - \varepsilon_y)\varepsilon_x(1 + K_i^d)]$$

$$= \frac{L}{u}[\varepsilon_z + (1 - \varepsilon_z)\varepsilon_y + (1 - \varepsilon_z)(1 - \varepsilon_y)\varepsilon_x + (1 - \varepsilon_z)(1 - \varepsilon_y)\varepsilon_x K_i^d]$$

$$= \frac{L}{v}[1 + \phi K_i^d] \tag{1}$$

where $$v = \frac{u}{\varepsilon_z + (1 - \varepsilon_z)\varepsilon_y + (1 - \varepsilon_z)(1 - \varepsilon_y)\varepsilon_x} \tag{2}$$

and $$\phi \equiv \frac{(1 - \varepsilon_z)(1 - \varepsilon_y)\varepsilon_x}{\varepsilon_z + (1 - \varepsilon_z)\varepsilon_y + (1 - \varepsilon_z)(1 - \varepsilon_y)\varepsilon_x} \tag{3}$$

The volumetric phase ratio for the lumped models is defined by Eq. (3). Eq. (2) relates the mobile phase or chromatographic velocity, v, to the superficial velocity, u.

The plate count (defined as the ratio of the square of the first moment to the second central moment) is given by [10]

$$\frac{1}{N_{plate}} = \frac{2}{Pe_z} + \frac{(1 - \varepsilon_z)\beta^2}{[\varepsilon_z + (1 - \varepsilon_z)\beta]^2}\left[\frac{2}{3}\frac{Pe_y\delta_y}{Sh} + \frac{1}{15}Pe_y\delta_y\right]$$

$$+ \frac{(1 - \varepsilon_z)}{(1 - \varepsilon_y)}\frac{(\beta - \varepsilon_y)^2}{[\varepsilon_z + (1 - \varepsilon_z)\beta]^2}\left(\frac{1}{15}Pe_x\delta_x\right) \tag{4}$$

where $$\beta \equiv \varepsilon_y + (1 - \varepsilon_y)\varepsilon_x(1 + K_i^d) \tag{5}$$

and the usual dimensionless terms (see List of Symbols) have been introduced. The plate count for each component can be calculated from Eq. (4) by substituting the appropriate dimensionless quantities; the component subscript is omitted for clarity.

When $(1 - \varepsilon_z)\beta \gg \varepsilon_z$ and $\beta \gg \varepsilon_y$, which are true for strongly-retained compounds, Eq. (4) reduces to the more familiar form [11]:

$$\frac{1}{N_{plate}} = \frac{2}{Pe_z} + \frac{2}{3}\frac{Pe_y\delta_y}{(1 - \varepsilon_z)Sh} + \frac{1}{15}\frac{Pe_y\delta_y}{(1 - \varepsilon_z)} + \frac{1}{15}\frac{Pe_x\delta_x}{(1 - \varepsilon_z)(1 - \varepsilon_y)} \tag{6}$$

It should be noted that Haynes and Sharma define their distribution coefficient, K^d, by

$$K^d = \frac{\rho S_x}{\varepsilon_x(1 - \varepsilon_y)}\frac{C_a}{C_x} \tag{7}$$

since their stationary phase concentration, C_a, is expressed in $gmol\,cm^{-2}$. Suitable modifications must be made if volumetric units are used instead. Further, Eq. (4) does not account for finite sorption rates, surface diffusion and bulk flow effects. If these contributions to bandspreading become significant in a given separation, the appropriate result for the plate count [12, 13] can be used in place of Eq. (4).

2.2 Craig Model

The Craig model is given by

$$c(i, j, k) + \phi q(i, j, k) = c(i, j - 1, k - 1) + \phi q(i, j, k - 1) \qquad (8)$$

where $c(i, j, k)$ is the mobile phase concentration of the ith species in the jth plate at the kth instant, $q(i, j, k)$ is the corresponding stationary phase concentration, and ϕ is the phase ratio. Using $q = K^d c$ and $k' = \phi K^d$, there follows

$$(1 + k_i')c(i, j, k) = c(i, j - 1, k - 1) + k'c(i, j, k - 1) \qquad (9)$$

Eq. (9) is a linear partial difference equation, whose solution can be found using standard methods [14, 15]. The analytical solutions to linear frontal chromatography and linear elution chromatography (when the pulse input fills several plates) are derived by the method of the two-dimensional z-transform in Appendix I; only the final forms are given here. The solution for linear elution is

$$\frac{c(i, j, k)}{c_0} = \begin{cases} (p_i')^j \sum\limits_{m=0}^{k-j} \begin{bmatrix} j + m - 1 \\ m \end{bmatrix} (q_i')^m, & k - j < M \\[2em] (p_i')^j \sum\limits_{m=k-j-M}^{k-j} \begin{bmatrix} j + m - 1 \\ m \end{bmatrix} (q_i')^m, & k - j \geq M \end{cases} \qquad (10)$$

where

$$p_i' = 1/(1 + k_i') \qquad (10a)$$

$$q_i' = k_i'/(1 + k_i') \qquad (10b)$$

and M is the input pulse width, and $\begin{bmatrix} n \\ r \end{bmatrix}$ is the binomial coefficient.

The analogous solution for frontal chromatography is

$$\frac{c_F(i, j, k)}{c_0} = (p_i')^j \sum_{m=0}^{k-j} \begin{bmatrix} j + m - 1 \\ m \end{bmatrix} (q_i')^m \qquad (11)$$

The effluent history (the experimentally obtained chromatogram) for a column of N_i plates is thus obtained by substituting $j = N_i$ in Eq. (10).

As is well known [16, 17], the Craig distribution results in a binomial distribution for the peak profile on-column. Equations (10) and (11) simply represent the finite sums of such binomial distributions, appropriate to inputs of a pulse of finite width and a step, respectively.

The results in Eqs. (10) and (11) can be formally generalized to non-integral J (plate count) and M (input pulse width) by making use of the identity

$$(p_i')^j \sum_{m=0}^{k-j} \binom{j + m - 1}{m} (q_i')^m = I_{p_i'}(j, k - j) \qquad (12)$$

where the right-hand-side of Eq. (12) is an incomplete beta function [18]. Thus the frontal solution can be represented as

$$\frac{c_F(i, N_i, k)}{c_0} = I_{p_i'}(N_i, k - N_i) \tag{13}$$

where p_i' is the probability of the ith species being in the mobile phase. The corresponding result for elution is

$$\frac{c(i, N_i, k)}{c_0} = I_{p_i'}(N_i, k - N_i) - I_{p_i'}(N_i, k - N_i - M) \tag{14}$$

Since the arguments of the incomplete beta function do not need to be integers, rational numbers may be used for N_i and M in Eqs. (13) and (14). Equation (12) is found in the theory of statistics [19]; an analogous result can be seen in the early work of Stene [20] on extraction.

The calculation of moments for the Craig distribution is also well established [21, 22]. These relations lead to a link between the plate count as defined earlier and the number of plates in the Craig column:

$$N_{plate, i} = N_i \left(\frac{1 + k_i'}{k_i'} \right) \tag{15}$$

Using Eq. (4) to calculate $N_{plate, i}$, the plate count of the ith component, and Eq. (15) allows the calculation of N_i, the number of divisions in a Craig column for which the same bandspreading will be produced. Setting $j = N_i$ in Eqs. (10) or (11) gives the corresponding analytical expression for the chromatogram.

The restriction to linear chromatography (required by Eq. (4)) guarantees that each component traverses the column independently of all others. Consequently, a plate count $N_{plate, i}$ can be determined for each component and the corresponding Craig plate number N_i determined. The analytical solutions for the multicomponent separation are then available without making the usual assumption that the components must be similar so that an average plate count can be used.

2.3 Continuous-Flow Plate Model

Also known as the "stirred-tank-in-series" model, this has also been widely used [17, 23–26]. The column is regarded as a series of vessels in each of which complete mixing and instantaneous equilibrium occurs. While the distance variable is thus discretized, time is retained as a continuous variable.

The mass balance within the jth vessel is

$$V_T \varepsilon_T \frac{dc(i, j, t)}{dt} + V_T (1 - \varepsilon_T) \frac{dq(i, j, t)}{dt} = F[c(i, j - 1, t) - c(i, j, t)] \tag{16}$$

where V_T is the volume of the vessel, ε_T the total porosity (as in Eqs. (1)–(3), this could be a combination of various porosities when compared to a complex rate model), and F the volumetric flow rate. The concentration notation is an extension of that previously used. As before this can be rewritten as

$$(1 + k_i') \frac{dc(i, j, t)}{dt} = \frac{v}{(\Delta x)} [c(i, j - 1, t) - c(i, j, t)]$$

$$= \frac{1}{(\Delta t)} [c(i, j - 1, t) - c(i, j, t)] \tag{17}$$

where v is the mobile-phase velocity, Δx ($= L/N$) is the length of the plate, and $v = \Delta x / \Delta t$.

Using $t_0 = L/v$ and $t_R = t_0(1 + k')$, there follows

$$\frac{dc(i, j, t)}{dt} = \frac{N_i}{t_{R,i}} [c(i, j - 1, t) - c(i, j, t)], \quad j = 1, 2, \ldots, N_i \tag{18}$$

The solution to the system represented by Eq. (18) is well known [e.g., 17]. For frontal chromatography with J plates, the chromatographic effluent history is given by

$$\frac{c_F(i, J, t)}{c_{i0}} = 1 - e^{-\alpha_i t} \sum_{j=0}^{J-1} \frac{(\alpha_i)^j t^j}{j!} = e^{-\alpha_i t} \sum_{j=J}^{\infty} \frac{(\alpha_i)^j t^j}{j!} \tag{19}$$

where $\alpha_i = N_i / t_{R,i}$. The corresponding elution profile, when the injection time of the pulse is τ, is given by

$$\frac{c(i, J, t)}{c_{i0}} = e^{-\alpha_i(t-\tau)} \sum_{j=0}^{J-1} \frac{(\alpha_i)^j (t-\tau)^j}{j!} - e^{-\alpha_i t} \sum_{j=0}^{J-1} \frac{(\alpha_i)^j t^j}{j!} \tag{20}$$

These analytical expressions are easy to use, and their moments (essentially those of the Poisson distribution) can be easily calculated [21]. Upon using these to evaluate the plate count, we get

$$J_i = N_{plate,i} \tag{21}$$

This equality is to be expected because the usual definition of plate count, Eq. (20), is in fact derived from continuous-flow plate theory [17]. It may be noted that the absence of the retention factor, in Eq. (21), together with the nature of the dependence of $N_{plate,i}$ on k_i' as seen in Eq. (4), imply that the continuous plate model will successfully predict the retention of an unretained solute.

These results for integral J can again be formally generalized as was done earlier for the Craig simulation. For all real J, the analog to Eq. (19) is

$$\frac{c_F(i, J, t)}{c_{i0}} = P\left(J_i, \frac{J_i t}{t_R}\right) \tag{22}$$

where the right-hand-side is one form of the incomplete gamma function [18]. Similarly, the result for elution is

$$\frac{c(i, J, t)}{c_{i0}} = P\left(J_i, \frac{J_i t}{t_R}\right) - P\left(J_i, J_i \left(\frac{t-\tau}{t_R}\right)\right) \tag{23}$$

An analogous result for extraction was again given by Stene [20].

2.4 Viability of the Craig Model

Ever since Glueckauf [24] stated in 1954 that the Craig, or discontinuous-flow, plate model as described by Mayer and Tompkins [27] overestimated the resolution achieved by a given column, its use has been limited. Glueckauf noted that the discrepancy is at all times quite significant but becomes extremely large for weakly-retained components: for example, the calculated number of theoretical plates is in error by over 100% for a component whose distribution coefficient is less than unity. However, we shall show below that when the plate count for the Craig model is appropriately defined, the Craig and continuous-flow plate models give very similar results. The only exception is an unretained component, for which, as is well known, the Craig model would give no bandspreading, i.e., the shape of the band is not altered at all by passage through the chromatographic column. This is obviously incorrect, since unretained components will also spread as a result of such processes as axial dispersion and pore diffusion, but this case is of limited practical interest.

First, we apply the results derived above to the case of frontal chromatography of one component through a single plate. This is obviously an unrealistic example, but serves to contrast the behavior of the two plate models, and can be generalized.

From Eq. (11), the effluent history for the Craig model is given by

$$\frac{c_F^{disc}(k)}{c_0} = p' \sum_{m=0}^{k-1} (q')^m \tag{24}$$

where the subscript i has been dropped, since only one component is being considered, and $j = 1$ (exactly one plate). This expression involves the sum of a geometric series; when this is evaluated, the result is

$$\frac{c_F^{disc}(k)}{c_0} = 1 - (q')^k = 1 - (1 - p')^k \tag{25}$$

The superscript "disc," standing for "discrete flow" or "discontinuous flow," has been added to distinguish the Craig from the continuous-flow model.

Under the same conditions, Eq. (19) gives

$$\frac{c_F^{cont}(t)}{c_0} = 1 - e^{-\alpha t} \tag{26}$$

where

$$\alpha = \frac{N}{t_R} = \frac{v}{(\Delta x)(1 + k')} = \frac{1}{(\Delta t)(1 + k')} \tag{27}$$

In order to compare this to the results from the Craig model, we consider the output at the finite values $t_k = k \, \Delta t$, for which

$$\frac{c_F^{cont}(k)}{c_0} = 1 - e^{-p'k} \tag{28}$$

Comparing the results for the two kinds of plate models, it is clear that $e^{-p'k} > 1 - p'$ and therefore $e^{-p'k} > (1 - p')^k$. Consequently $c_F^{disc}(k) > c_p^{cont}(k)$, for all $t = t_k$. Thus the Craig effluent is always higher than that from the continuous-flow plate, will reach the initial value faster, and is therefore more efficient. These results are exemplified in Fig. 1, for $k' = 1$. Since the argument will extend to any finite number of plates, it is clear that a column composed of a certain number of Craig plates will be more efficient than one composed of an equal number of continuous-flow plates. Thus, when examining experimental data, one must distinguish between descriptions based on Craig plates and continuous-flow plates. In fact, it is better to avoid the term "plate" and speak of Craig segments and continuous-flow segments (these latter might well be called Glueckauf segments). Then no confusion will arise with the experimentally well-defined plate number:

$$N_{plate} = \left(\frac{t_R}{\sigma_t}\right)^2 \tag{29}$$

for the effluent peak of any component where t_R is its retention time (the time of emergence of its center of mass) and σ_t is its standard deviation (in consistent time units).

The connection between the experimentally determined plate count and the number of plate segments needed to generate exactly this degree of bandspreading has been summarized by Fritz and Scott [21] given earlier as Eqs. (15)

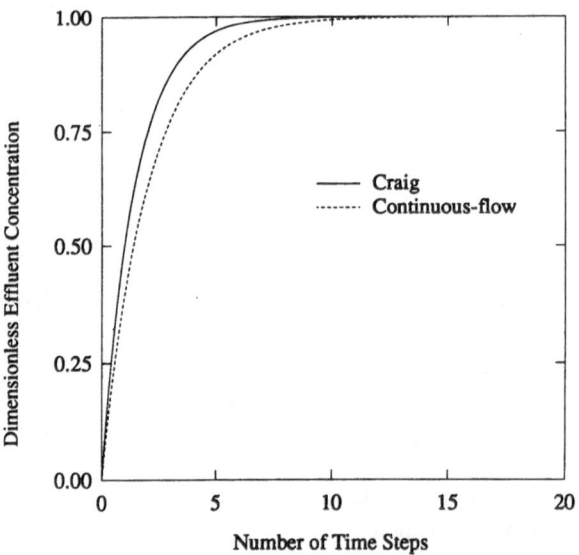

Fig. 1. Comparison of effluent histories from a single Craig and continuous-flow segment for frontal chromatography (adsorption)

and (21). They are repeated here in the present notation for the purposes of comparison.

$$N^{cont} = N_{plate} \tag{30}$$

$$N^{disc} = N_{plate}\left(\frac{k'}{1 + k'}\right) \tag{31}$$

Thus fewer Craig than Glueckauf segments are needed to generate a certain dispersion; this is consistent with our earlier observation that Craig segments are the more efficient. This also explains the discrepancy described by Glueckauf: when $k' \sim 1$, the difference between N^{disc} and N^{cont} is on the order of 100%. It can also be seen from Eq. (31) that when $k' = 0$, $N^{disc} = 0$, implying that the Craig model cannot capture the bandspreading of an unretained component. With this exception, however, the Craig model is a useful tool.

Just as the Glueckauf segment can be regarded as a continuous stirred-tank adsorber (CSTA), there is a connection between the Craig segment and a plug flow adsorber (PFA). If we imagine a PFA of length Δx packed with adsorbent into which material is fed from the preceding PFA such that it takes a time Δt to completely fill it, the discretized mass balance is

$$c(x - \Delta x, t - \Delta t) + \phi q(x, t - \Delta t) = c(x, t) + \phi q(x, t) \tag{32}$$

which is exactly that of the Craig segment.

This also explains the lack of bandspreading in a Craig segment for $k' = 0$: the PFA by definition possesses no bandspreading effects in the absence of adsorption. It is only by virtue of finite retention that a series of PFAs, or Craig segments, will generate bandspreading. This is in contrast to the Glueckauf segment wherein bandspreading occurs by mixing over the entire volume of the segment, even in the absence of retention.

3 Results and Discussion

Either plate model described here would seem to be a simple, if approximate, alternative to the detailed rate models. In choosing between them, the fact that the Craig distribution does not predict the dispersion of an unretained solute could argue against it. On the other hand, in the examples given below, this model performs at least as well as the continuous-flow plate model when $k' \neq 0$.

One disadvantage of plate models is that they cannot satisfy the Danckwerts boundary conditions [28]. However, it is known that the choice of boundary conditions is only significant when the axial Peclet number is small and the column is relatively short [11]: a combination that occurs infrequently in chromatographic practice. The examples show how the choice of model is largely a matter of convenience.

3.1 Use of Plate Models to Recover Solutions of Rate Models

Rasmuson has derived analytical solutions for detailed rate models of linear chromatography [29, 30]. However, these solutions are complex and involve infinite integrals of oscillatory arguments; simpler approaches that retain sufficient accuracy would be useful in design and scale-up. Here, simulations based on the Craig and continuous-plate models are compared to results from the literature. Theoretical or numerical results were used instead of experimental data in order to assess the difference between the results of the simple and complex models while avoiding the additional scatter that is inherent in experimental data.

Detailed simulations of a complex rate model describing the sorption of a single component on a bidisperse sorbent have been carried out by Haynes and Sharma [8] and Raghavan and Ruthven [31]. Axial dispersion, film mass-transfer, and micro- and macro-pore diffusion are accounted for. A representative result from Raghavan and Ruthven [31] was chosen for comparison to the plate models.

In order to carry out the Craig simulation, the column data (such as dimensions, and values of the various porosities) are used to calculate the volumetric phase ratio from Eq. (3). Since the distribution coefficient K^d is given, the retention factor k' can be calculated $(= \phi K^d)$. The probabilities p' and q' can then be found from Eqs. (10a) and (10b). The chromatographic velocity can be calculated from Eq. (4); this, together with the column length, specifies the column hold-up time. The plate count, N_{plate}, can be calculated from Eq. (4). Sometimes data is reported in terms of dimensionless groups other than those used in Eq. (4), but only trivial algebra is needed to calculate the required terms. The number of Craig segments is then calculated from Eq. (31). The (discrete) input pulse width is the final parameter needed. In this case, the effluent history is in dimensionless time units, being scaled to the column hold-up time [31]; then the input pulse width must also be scaled. The concentration is also reported in dimensionless terms, scaled to its input inlet value. With this information, the Craig simulation as expressed by Eq. (14) can be carried out. A FORTRAN program embodying this calculation is given in Appendix II (the parameter values in the program correspond to a later simulation, shown in Fig. 5). The IMSL function for the incomplete beta function is used. This detailed description of how a representative plate simulation was carried out is given only to emphasize the simplicity of the method.

The continuous-flow plate simulation is approached in the same way. The retention time t_R (i.e., the first moment) is calculated from Eq. (1). The input pulse width τ (in appropriate units) can be used directly, since time is a continuous variable in this method. The number of continuous-flow segments is calculated from Eq. (30). These parameters are then used in either Eq. (20) or (23).

The results of these simulations were compared to that of Raghavan and Ruthven [31] in Fig. 2; the agreement is excellent.

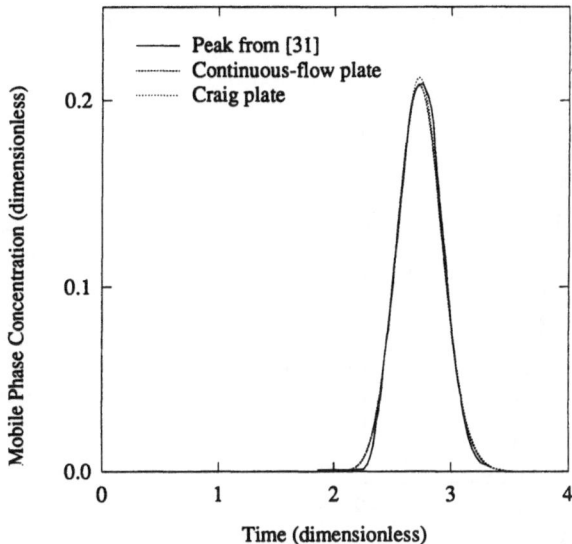

Fig. 2. Linear elution. Comparison of plate simulations with the numerical results of Raghavan and Ruthven [31] for linear elution. The parameter values are: $\varepsilon_x = 0.42$, $\varepsilon_y = 0.32$, $\varepsilon_z = 0.41$; $K^d = 11.4$; $Pe_z = \infty$; $Sh = 2,000$; $\delta_y Pe_v = 3.66 \times 10^{-2}$; $Pe_x \delta_x = 5.06 \times 10^{-3}$; $N_{plate} = 200.1$

3.1.1 Effect of Sample Volume

Another example of elution chromatography is used to examine the effect of sample volume. Carta [32] used a rate model where pore diffusion was the dominant bandspreading mechanism. His analytical solution for two sample volumes is compared with the results from the plate models in Fig. 3 (a and b). Since a monodisperse pore distribution is assumed, the appropriate retention time equation is not Eq. (1), but the following:

$$\mu_{1,i} = \frac{L}{u}[\varepsilon_z + (1 - \varepsilon_z)\varepsilon_y + (1 - \varepsilon_z)(1 - \varepsilon_y)K_i^d]$$

$$= \frac{L}{u}[\varepsilon_z + (1 - \varepsilon_z)\{\varepsilon_y + (1 - \varepsilon_y)K_i^d\}]$$

$$= \frac{L}{v_{int}}\left[1 + \frac{1 - \varepsilon_z}{\varepsilon_z}K_i^*\right] \tag{33}$$

where $K_i^* = \varepsilon_y + (1 - \varepsilon_y)K_i^d$ and $v_{int} = \dfrac{u}{\varepsilon_z}$.

Carta only reported the lumped value K_i^* since his system involved a gel, which is unlikely to possess a distinct macroporosity. An analogous form of Eq. (4) is then used to calculate the plate count, with $Sh \to \infty$ [11]. The plate simulations can then be carried out as before.

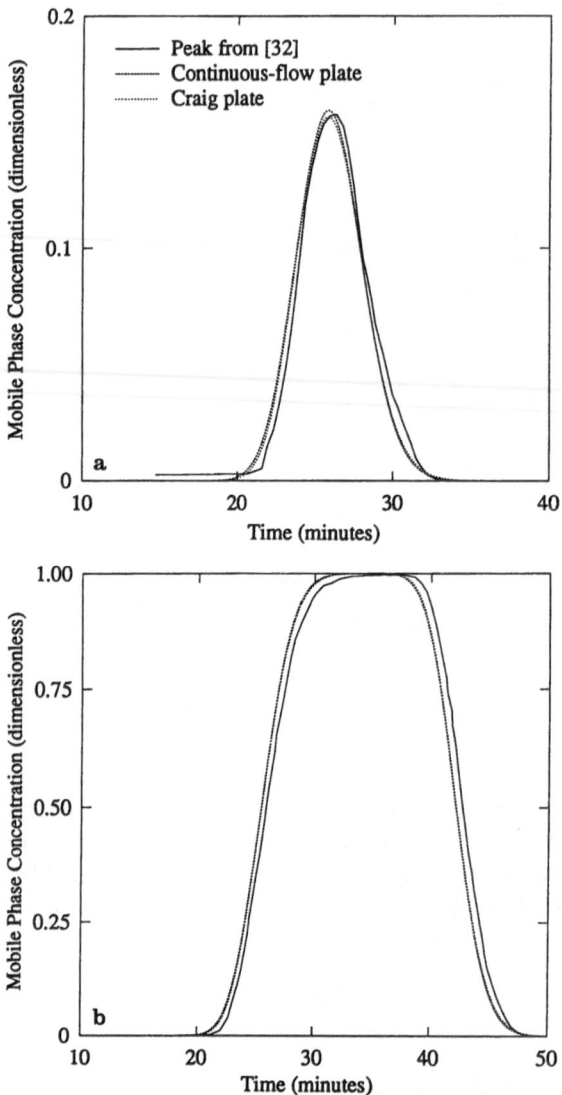

Fig. 3a and b. Linear elution of fructose with large inputs (volume-overloaded elution). Plate simulations are compared to the numerical results of Carta [32]. $\varepsilon_z = 0.39$; $Sh = 892.9$; $\delta_y Pe_y = 0.231$; $K^* = 0.66$; $N_{plate} = 152.1$. Feed pulse duration is **a** 50 s and **b** 1000 s. Symbols as in Fig. 2

Again, as can be seen from Fig. 3, the plate models agree well with the more complex rate model. Note that the contribution of the finite widths of the input pulse to the chromatogram (its contribution to the first and second moments is well known, e.g., Sternberg [33]) is automatically accounted for by the solutions used here; no additional correction is needed.

3.1.2 Glucose/Fructose Separation

Since each component can be assigned its own plate number in linear chromatography, plate models can also be used to accurately describe multi-component separations. Fig. 4 shows the comparison of plate simulations with Carta's results on the separation of fructose from glucose [32]. The retention time and plate count were calculated as in Fig. 3. Good agreement is achieved. It may be noted in Fig. 4 that the fructose peak from the rate model [32] does not begin from zero concentration. This is because the rate model was used to model a cyclic separation process; only one cycle was used here in the comparison with the plate models.

3.2 Generalizations of Plate Models

Plate models ensure accurate regeneration of the first and second, but not necessarily of the higher, moments. Skewed peaks (characterized by non-zero third central moments), can be generated from plate models when the plate count is very low ($N_{plate} < 20$), but not at the significantly higher plate counts usually found in actual separations. Since skewed peaks emerging from efficient columns tend to indicate nonlinear, rather than linear, chromatography,

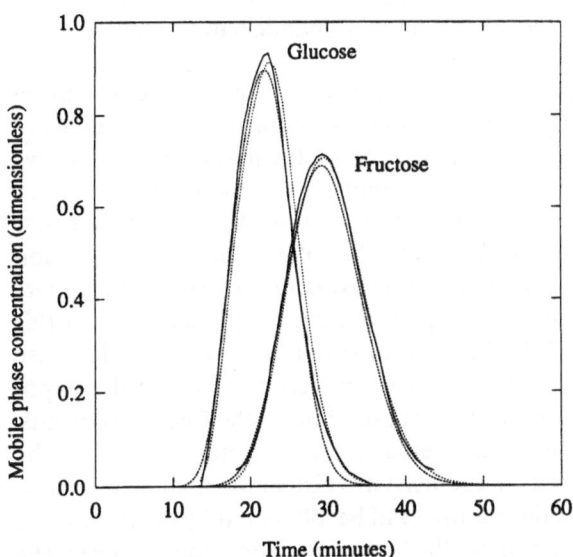

Fig. 4. Multicomponent linear elution. Comparison of plate simulations with the numerical results of Carta [32]. $\varepsilon_z = 0.39$. For glucose, Sh $= 4,545.5$; $\delta_y Pe_y = 2.35$; $K^* = 0.26$; $N_{plate} = 46.5$. For fructose, Sh $= 1,785.7$; $\delta_y Pe_y = 0.92$; $K^* = 0.66$; $N_{plate} = 38.2$. Symbols as in Fig. 2

the usefulness of plate models in simulating nonlinear separations is also examined.

3.2.1 Nonlinear Chromatography

The assumption that each component traverses the column independently of the others allowed the effluent history of each component to be calculated separately. Carrying out large-scale separations in the linear mode can be useful in practice when the sorption isotherms stay linear until a relatively high mobile phase concentration, as is frequently the case with sugars. There are, however, some practically important chromatographic modes for which this assumption is no longer valid.

One example is mutual interference caused by high concentrations of the feed components. Here the movement of one component depends on the presence and concentrations of the others, and a common plate count must be used. Such simulations have been widely used by Guiochon and co-workers [e.g., 34, 35] and Snyder and co-workers [e.g. 36, 37]. The usual justification is that the bandspreading suffered by all the feed components is comparable, an assumption which is frequently reasonable for chromatographic separations. In addition, the curvature of the (multicomponent) sorption isotherms also gives rise to bandspreading, since different concentrations will travel with different velocities, and this "thermodynamic contribution to bandspreading" could be a substantial fraction of the total bandspreading. Under these circumstances, plate simulations could still give reasonably accurate results. (The plate models do not have to be extended to account for thermodynamic bandspreading, since it is "built into" the isotherms themselves.)

Another practical instance of interference arises in gradient elution chromatography. Even when the feed components are in the linear regions of their own sorption isotherms, they are influenced by the mobile phase additive (such as salt in ion-exchange chromatography and organic modifier in reversed-phase chromatography) which modulates their retention. Thus the plate count of each feed component varies on moving down the column, since their retention factor varies [38, 39]. It is still possible to approximate the process by a plate simulation, such as the continuous-flow plate model used extensively in this context by Yamamoto et al. [26]. Here the distribution of the mobile phase additive is first solved for (assuming that it is unaffected by the feed components), and this solution is used in the plate simulations of the feed components, where the plate count appropriate to an averaged retention factor is used. In the more complex case where the feed components are also in the nonlinear regions of their own sorption isotherms, the additive will be influenced by the feed, and it would become preferable to simulate the complete rate model. When the modulator is in the nonlinear portion of its own isotherm, peak shapes can be dramatically affected [40]. However, if modulator adsorption is appropriately accounted for, plate simulations can still capture the key features of the chromatogram.

3.2.2 Non-Equilibrium Phenomena

Plate models can be generalized to describe specific features. The classical papers by Deans and Lapidus [23] used a two-dimensional network of plates to describe radial as well as axial dispersion. Returning to the one-space-dimensional description, Villermaux [41] considered a version of the tanks-in-series model in which mass transfer occurs between the mobile and stationary phases (in the original form, instantaneous equilibrium is assumed in each tank). The result was

$$\frac{1}{N_{plate}} = \frac{1}{N} + \frac{2v}{L} \frac{1}{k_{MT}} \frac{k'}{(1 + k')^2} \tag{34}$$

The plate count from the rate model incorporating axial dispersion and an overall mass transfer coefficient [11] is:

$$\frac{1}{N_{plate}} = \frac{2D}{vL} + \frac{2v}{L} \frac{1}{k_{overall}} \frac{k'}{(1 + k')^2} \tag{35}$$

On comparing Eqs. (34) and (35), it can be seen that the mass transfer coefficient k_{MT} in the plate model can be directly equated to the overall mass transfer coefficient. This in turn can be set equal to the sum of the individual contributions from film mass transfer, micro- and macro-pore diffusion [8]. The finite nature of the plates must then generate a bandspreading equal only to that produced by axial dispersion in the rate model: $N = \frac{vL}{2D}$ where D is the axial dispersion coefficient. This result is to be expected from the rate model, where the mass-transfer and the dispersion contributions to bandspreading are separated. It is an indication of the versatility of plate models that a similar separation is found in Eq. (35).

An additional advantage of plate models is their simplicity in describing the kinetic terms which allow the incorporation of complex equilibrium behavior without resulting in an intractably complex simulation. This has been used to advantage by Wankat [42], who analyzed the interaction of an enzyme with various other components in affinity chromatography. In fact, given that the dominant kinetic contribution in affinity separations is usually slow adsorption-desorption kinetics, it might be appropriate to use an analog of the Villermaux plate model, in which an explicit rate equation for the binding of the substrate is added to the usual mass balance.

Thus, plate models, when used with the appropriate plate height expression, provide a simple and reasonably accurate approach to modeling fixed-bed sorption. Even when a detailed description of the process is required, e.g., in the design of the purification of a high-valued product, such plate simulations could be used as an initial approximation to, and could provide valuable information for, more complex simulations of the coupled PDE's that govern the process.

3.3 Application of Plate Models to Scale-Up

An important advantage of the plate models for linear chromatography is in the relative ease with which they can be applied to scale-up. The rigorous analytical solutions to the complete rate models that have been previously mentioned [29, 30] are complex, and involve infinite integrals with oscillatory arguments. The analytical solutions available for plate models are exact, explicit, and quite simple in form. The use of these solutions scaling separations is outlined next.

The explicit solutions to the Craig model – Eqs. (10) and (11), or (13) and (14) – and the continuous-flow model – Eqs. (19) and (20), or (22) and (23) – can be coupled with the appropriate expression for the plate count, e.g., Eq. (6), to relate the chromatographic effluent histories to the various parameters such as the mass-transfer coefficient and the pore diffusivity that determine bandspreading.

The design parameters involved in scale-up are the length and diameter of the column, the particle size (assuming a monodisperse distribution of particles), the volumetric flow rate (or equivalently the mobile phase velocity) and the sample size. The sample composition is usually given. External constraints include the pressure drop and the effluent purity (or, equivalently, the chromatographic resolution). The parameter to be optimized is the throughput or production rate, which are measures of the amounts of acceptably pure material generated per unit time. Mathematically, this can be regarded as a problem in constrained optimization; the explicit dependence in plate solutions of the optimization function on the design parameters should permit rapid numerical solution.

However, in most cases the separation has already been successfully developed at bench-scale. The objective is then to scale the operating parameters such as particle and column size appropriately so as to produce the same separation at preparative-scale. The specification of design parameters on this basis has been studied by several workers [43, 44–46]. The plate models facilitate ready evaluation of such design recommendations.

Figure 5 shows a comparison of large-scale isocratic elution data for the separation of phenyl alanine from aspartame taken from Ladisch et al. [47] to results from the Craig model. The various operating parameters, taken from [46], are given in the caption to Fig. 5. The dimensionless parameters representing the contributions of film mass transfer, pore diffusion, etc., are calculated using standard correlations in the literature. The Sherwood number is calculated using the Wilson and Geankoplis correlation [48]. The adsorbate's molecular diffusivity and pore diffusivity are calculated as in Ladisch et al. [47] except that a tortuosity factor of 3 was used. The parameters needed for the Craig simulation were then calculated as described earlier. Thus, a plate calculation involving no fitted parameters is able to achieve reasonable agreement with experiment, as can be seen in Fig. 5. It should be noted that the standard deviation associated with the particle size was 4.9 μm (the particle size itself is 60.3 μm). This size distribution could result in a wider band than would be predicted by the plate model.

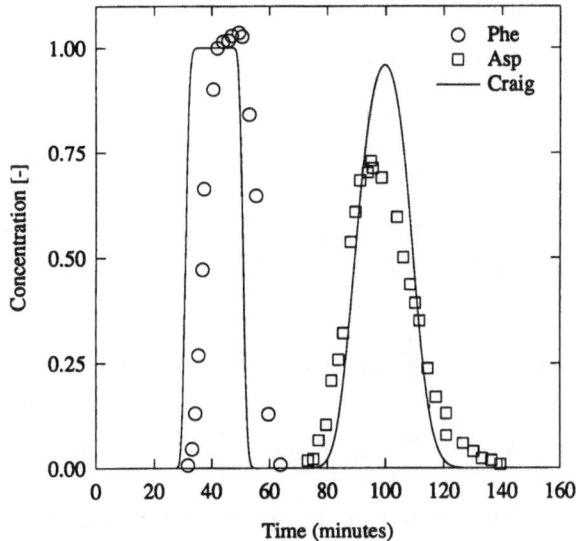

Fig. 5. Scaled-up linear isocratic elution separation of phenyl alanine from aspartame; experimental data from Ladisch et al. [47]. Column dimensions: 1.09×70 cm; operating temperature: 20 °C; flow rate = 2 ml min^{-1}; interstitial porosity = 0.36; total porosity = 0.74; feed: 40 ml of 5 mg ml^{-1} of both components; particle size = 60.3 µm. Symbols: ○, phenyl alanine concentrations from [47]; □, aspartame concentrations from [47]; ———, Craig simulation

4 Conclusions

Plate models are shown to be simple and accurate approximations to detailed rate models in linear chromatography. Regarding a single Craig plate as a PFA explains its lack of bandspread for an unretained solute, and emphasizes its similarity to a continuous-flow plate (PFA). Both models are shown to be useful in predicting multicomponent separations. Possible application to nonlinear chromatography is discussed. The analytical solutions available for the Craig and continuous-flow plate models are attractive in the scale-up and design of preparative separations.

Acknowledgement. This work was supported by NSF grants CBT8351916 and BCS8912150. We thank Professor P. Wankat and Professor G. Narasimhan for their helpful comments.

5 Appendix I: Solutions to the Craig Model

The basic Craig description (Eq. (8) in the text) is

$$c(i, j, k) + \phi q(i, j, k) = c(i, j - 1, k - 1) + \phi q(i, j - 1, k - 1) \qquad \text{(A1-1)}$$

which can be rewritten for linear chromatography as

$$c(i, j, k) = p_i' c(i, j - 1, k - 1) + q_i' c(i, j, k - 1) \qquad (A1\text{-}2)$$

where

$$p_i' \equiv \frac{1}{1 + k_i'}, \quad q_i' \equiv \frac{k_i'}{1 + k_i'}, \quad p_i' + q_i' = 1 \qquad (A1\text{-}3)$$

This linear partial difference equation can be solved by several methods [15]. Here we use the two-dimensional z-transform [14]. If the z-transform of $c(i, j, k)$ is $U(i, z, w)$, Eq. (A1-2) becomes

$$zw \left[U(i, z, w) - \sum_{j=1}^{\infty} c(i, j, 0)z^{-j} - \sum_{m=1}^{\infty} c(i, 0, k)w^{-m} + c(i, 0, 0) \right]$$

$$= p_i' U(i, z, w) + q_i' z \left[U(i, z, w) - \sum_{m=1}^{\infty} c(i, 0, k)w^{-m} \right] \qquad (A1\text{-}4)$$

Letting $c(i, j, k)$ be dimensionless (through division by the input concentration c_{i0}), the boundary condition for elution chromatography with a finite pulse is

$$c(i, 0, k) = \begin{cases} 1, & 0 \le k \le M \\ 0, & k > M \end{cases} \qquad (A1\text{-}5)$$

where M represents the (discretized) injection time. The initial condition corresponding to an empty column that agrees with Eq. (A1-5) is

$$c(i, j, 0) = \begin{cases} 1, & j = 0 \\ 0, & j \ge 1 \end{cases} \qquad (A1\text{-}6)$$

For frontal chromatography, Eq. (A1-6) remains valid, and the boundary condition is

$$c(i, 0, k) = 1 \quad \text{for all } k \ge 0 \qquad (A1\text{-}7)$$

Thus, letting $M \to \infty$ in Eq. (A1-5) will give the result for frontal chromatography. Substituting Eqs. (A1-5) and (A1-6) into (A1-4), we have, for elution chromatography,

$$U(i, z, w) = \frac{(zw - q_i' z)}{(zw - q_i' z - p_i')} \sum_{m=0}^{M} w^{-m}$$

$$= \frac{1}{1 - \dfrac{p_i'}{z(w - q_i')}} \sum_{m=0}^{M} w^{-m} \qquad (A1\text{-}8)$$

This can be inverted with respect to z to get

$$\tilde{c}(i, j, w) = \left(\frac{p_i'}{w - q_i'} \right)^j \sum_{m=0}^{M} w^{-m} \qquad (A1\text{-}9)$$

We note that m is a dummy summation variable, and is analogous to a dummy integration variable. To invert with respect to w, we rewrite Eq. (A1-9) as

$$\tilde{c}(i, j, w) = (p_i')^j w^{-j} \left(1 - \frac{q_i'}{w}\right)^{-j} \sum_{m=0}^{M} w^{-m} \tag{A1-10}$$

Inversion can now be carried out to obtain, for $k - j \geq M$,

$$c(i, j, k) = (p_i')^j \sum_{m=k-j-M}^{k-j} (-1)^m \begin{bmatrix} -j \\ m \end{bmatrix} (q_i')^m \tag{A1-11}$$

where $\begin{bmatrix} a \\ b \end{bmatrix}$ is the binomial coefficient, given for $a \geq b$ by $\begin{bmatrix} a \\ b \end{bmatrix} = \dfrac{a!}{b!(a-b)!}$.

Using a relation for negative indices in a binomial coefficient [49], there follows

$$c(i, j, k) = (p_i')^j \sum_{m=k-j-M}^{k-j} \begin{bmatrix} j + m - 1 \\ m \end{bmatrix} (q_i')^m \tag{A1-12}$$

For $k - j < M$, the solution is

$$c(i, j, k) = (p_i')^j \sum_{m=0}^{k-j} \begin{bmatrix} j + m - 1 \\ m \end{bmatrix} (q_i')^m$$

It can be seen that Eq. (A1-12) represents a finite sum of binomial expressions, as might be expected for a finite pulse input. The corresponding result for frontal chromatography is easily obtained from Eq. (A1-12) by letting $M \to \infty$:

$$c_F(i, j, k) = (p_i')^j \sum_{m=0}^{-j+k} \begin{bmatrix} j + m - 1 \\ m \end{bmatrix} (q_i')^m \tag{A-14}$$

Eq. (A1-12) and (A1-14) are given in the text as Eq. (10) and (11), where the concentrations have been returned to dimensional form.

6 Appendix II: Computer Program Based on the Craig Model

```
            implicit real * 8 (a–h, o–z)
            real * 8 kprime

            data imode /2/
c           imode = 1 for frontal, = 2 for elution chromatography

            data t0, time_in, kprime, plate /24.2, 202.6, 2.7, 245.0/
c           t0:retention time of an unretained component that fully
c               explores the mobile phase space (i.e., no size exclusion occurs)
c           time_in:the feed volume, in time units
c           kprime:the retention factor of the adsorbate
c           plate:the number of Craig segments (need not be an integer)
```

```fortran
      open (unit = 8, file = 'craig3 .mass', status = 'unknown')

      delt = t0/plate
      p = 1.0d0/(1.0d0 + kprime)

      kstart = ifix (plate) + 1
      kend = plate + 1250
      arg1 = dfloat (plate)
      amount = 0.0d0
      do 50 k = kstart, kend

         time = delt * dfloat (k)
         arg2 = dfloat (k) – plate
         out1 = dbetai (p, arg1, arg2)
         if (mode .eq. 1) amount = amount + out1

         if (imode .eq. 2) then
            arg3 = dfloat (k) – time_in – plate
            if (ifix(arg3) .gt. 0) then
               out2 = dbetai (p, arg1, arg3)
            else
               out2 = 0.0d0
            end if
            out1 = out1 – out2
            amount = amount + out1
         end if

         if ((imode .eq. 1) .and. (dabs(out1 – 1.0d0) .gt. 1.0d–6)
     $        .and. (out1 .gt. 1.0d–6)) write (6,1000) k, time, out1
         if ((imode .eq. 2) .and. (out1 .gt. 1.0d–6))
     $        write (6,1000) k, time, out1

         n_out = mod (k, 100)
         if (n_out .eq. 0) write (8, *) 'k =', k, ', amount =',
     $                                  amount
50       continue
         stop
1000     format (1x, i5, 2x, g12.5, 3(2x, g20.13))
         end
```

7 References

1. Giddings JC (1965) Dynamics of chromatography. Marcel Dekker, New York
2. Yang C-M, Tsao GT (1982) In: Fiechter A (ed) Advances in biochemical engineering, vol 25. Springer, Berlin Heidelberg New York, p 1
3. Giddings JC (1965) Dynamics of Chromatography. Marcel Dekker, New York
4. Horvath Cs, Lin H-J (1978) J Chromatogr 149: 43
5. Klinkenberg A, Sjenitzer F (1956) Chem Eng Sci 5: 258
6. Kucera E (1965) J Chromatogr 19: 237
7. van Deemter JJ, Zuiderweg FT, Klinkenberg A (1956) Chem Eng Sci 5: 271
8. Haynes HW, Sharma PN (1973) AIChE J 19: 1043
9. Kubin M (1965) Coll Czech Chem Commun 30: 1104
10. Haynes HW, Jr (1975) Chem Eng Sci 30: 955
11. Ruthven DM (1984) Principles of adsorption and adsorption processes. John Wiley, New York
12. Furusawa TM, Suzuki M, Smith JM (1976) Catal Rev-Sci Eng 13: 43
13. Weber TW, Chakravorti RK (1974) AIChE J 20: 228
14. Jury EI (1964) Theory and application of the z-transform method. John Wiley, New Jersey
15. Uspensky JV (1937) Introduction to mathematical probability. McGraw-Hill, New York
16. Karger BL, Snyder LR, Horvath CS (1973) An introduction to separation science. John Wiley, New York
17. Keulemans AIM (1957) Gas chromatography, Reinhold, New York
18. Abramowitz M, Stegun I (1965) Handbook of mathematical functions with formulas, graphs, and mathematical tables. Dover, New York
19. Pearson K (1934) Tables of the incomplete beta function. Biometrika Office, London
20. Stene S (1945) Arkiv Kemi 18A(18): 1
21. Fritz JS, Scott DM (1983) J Chromatogr 271: 193
22. Mood AM, Graybill FA (1963) Introduction to the theory of statistics. McGraw-Hill, New York
23. Deans HA, Lapidus L (1960) AIChE J 6: 656
24. Glueckauf E (1955) Trans Faraday Soc 51: 34
25. Said AS (1956) AIChE J 2: 477
26. Yamamoto S, Nakanishi K, Matsuro R (1988) Ion-exchange chromatography of proteins. Marcel Dekker, New York
27. Mayer SW, Tompkins ER (1947) J Am Chem Soc 69: 2866
28. Danckwerts PV (1953) Chem Eng Sci 2: 1
29. Rasmuson A (1982) Chem Eng Sci 37: 787
30. Rasmuson A, Neretnieks I (1980) AIChE J 26: 686
31. Raghavan NS, Ruthven DM (1985) Chem Eng Sci 40: 699
32. Carta G (1988) Chem Eng Sci 43: 2877
33. Sternberg JC (1966) Adv Chromatogr 2: 205
34. Czok M, Guiochon G (1990) Anal Chem 62: 189
35. Lin B, Guiochon G (1989) Sep Sci Tech 24: 31
36. Eble JE, Grob RL, Antle PE, Cox GB, Snyder LR (1987) J Chromatogr 405: 31
37. Snyder LR, Cox GB, Antle PE (1988) J Chromatogr 444: 303
38. Jandera P, Churacek J (1985) Gradient elution in column liquid chromatography. Elsevier, Amsterdam
39. Poppe H, Paanakker J, Bronckhorst M (1981) J Chromatogr 204: 77
40. Velayudhan A, Ladisch MR (1991) Anal Chem 63: 2028
41. Villermaux J (1981) In: Rodrigues AE, Tondeur D (eds) Percolation processes. Sijthoff and Noordholt, Rockville
42. Wankat PC (1974) Anal Chem 46: 1400
43. Snyder LR (1972) J Chromatogr Sci 10: 369
44. Wankat PC, Koo Y-M (1988) AIChE J 34: 1006
45. Ladisch MR, Wankat PC (1988) In: Phillips M, Shoemaker SP, Middlekauff RD, Otterbrite M (eds) Impact of chemistry on biotechnology, ACS Symposium Series, No. 362, American Chemical Society, Washington, DC
46. Lee CK, Yu Q, Kim SU, Wang N-HL (1989) J Chromatogr 484: 29
47. Ladisch MR, Hendrickson RL, Firouztale E (1991) J Chromatogr 540: 85
48. Wilson EJ, Geankoplis CJ (1966) Ind Eng Chem Fundam 5: 9
49. Knuth DE (1969) The Art of computer programming. Addison-Wesley, Reading

Liquid Chromatography Using Cellulosic Continuous Stationary Phases

Yiqi Yang[1,*], Ajoy Velayudhan[2], Christine M. Ladisch[1],
and Michael R. Ladisch[2,3]
[1] Department of Consumer Sciences and Retailing;
[2] Laboratory of Renewable Resources Engineering;
[3] Department of Agricultural Engineering,
Purdue University, West Lafayette, IN 47907, USA

A novel type of continuous stationary phase based on fabric materials is described. This column packing utilizes the continuous character of a cellulose (cotton) stationary phase, and the chemistry of the derivatized forms of the adsorbent, to obtain separations of proteins and small molecules based on cation and anion exchange, hydrophobic interactions, and size. The mechanical stability of the stationary phase facilitates chromatographic velocities in excess of 70 cm min^{-1}. The influence of eluent properties on the adsorption of sample proteins is discussed in this chapter. Sequential stepwise desorption is used to separate 100 µl mixtures of BSA, IgG, β-galactosidase, and insulin in 10 minutes or less, using 10 mm i.d. × 500 mm length columns.

* Current address: Division of Consumer Sciences, University of Illinois at Urbana-Champaign, Urbana, IL 61801

Advances in Biochemical Engineering
Biotechnology, Vol. 49
Managing Editor: A. Fiechter
© Springer-Verlag Berlin Heidelberg 1993

1 Introduction

Separation costs in the manufacture of proteins and other biotechnology products are estimated to be 40%–75% of the production cost [1]. Since preparative chromatography is an important component in the downstream processing of proteins, ways need to be found to make large-scale chromatographic separations more efficient.

Key factors which improve the cost-effectiveness of a chromatographic process are reduction of the number of steps, automation to reduce labor costs, and reduction of media costs [2]. Changes in upstream processing to obtain a less heterogeneous product stream can simplify recovery and purification, thereby reducing costs. However, improvements in separations technology are still needed, particularly in reducing residence time. Chromatographic techniques favored by both producers and regulatory agencies are ion exchange, hydrophobic interaction, affinity, and size-exclusion chromatography [3].

On an analytical scale, small sample volumes and low solute concentrations in relatively clean samples are preferred, since these give the highest resolution. On a process scale, high throughput and short residence times with attainment of extraordinary product purity are key objectives. Special stationary phases designed specifically for this purpose seem appropriate, if not essential. Whether an analytical-scale separation can even provide useful information for the design of the corresponding large-scale separation under conditions of mass overload (when throughputs are high) has been questioned [4].

A successful process-scale system must also provide high mass transfer rates at low pressure drop; this is equivalent to a high ratio of mass to momentum transfer. In this context, Gibbs and Lightfoot [5] suggested hollow fibers as possible chromatographic stationary phases, although they believe that problems of fabrication and flow distribution still remained to be solved. In 1989, Ding et al. [6], expanded on Gibbs and Lightfoot's work, and successfully used a bundle of hollow fibers. They coated the fibers with dodecanol and trioctylphosphate/dodecane and used these systems in chromatographic columns to separate proteins. Eluents with pH's as high as 10–11 were used in order to accelerate elution; separation times of 30–80 min were obtained.

This work uses another type of monodisperse cylindrical packing: whole fabrics made of yarns whose diameters range from 100–150 μm. These materials (diethyl-aminoethyl-cotton, sulfated cotton, and viscose rayon), when rolled into a cylinder and inserted into a metal column, can be used as stationary phases. It should be emphasized that a single block of fabric constitutes the stationary phase. Hence the terms "continuous stationary phase" or "rolled stationary phase" (RSP) are used, to distinguish the present sorbent from conventional particulate sorbents. Such columns, which can be run like conventional LC columns, were mechanically stable at chromatographic velocities in excess of 70 cm min^{-1} and separated insulin from β-galactosidase (β-gal) in

4 min, and an immunoglobulin G (IgG) from bovine serum albumin (BSA) in 3 min. A mixture of these four proteins was completely separated in less than 10 min.

This chapter describes the experimental results for RSP columns of cation, anion, and hydrophobic interaction stationary phases from polyhydroxyl fabrics, all of which gave baseline separations in less than 10 min. Preliminary characterization of these columns with respect to eluent velocity, salt and pH gradients indicated the stationary phases were durable and mechanically stable. Further work is now needed to examine loading and sample concentration effects. Design of large-scale systems will also require that models of mass transfer and diffusional phenomena in rolled stationary phases be developed and correlated to the well-known chromatographic parameters usually associated with packed beds of spherical particles, or with the more recently reported hollow fiber models.

2 Background

Particulate supports are the most common kind of chromatographic stationary phase currently in use. However, fibers, yarns and fabrics can also be used to pack liquid chromatography columns, and can have several advantages. The polymers which form fibers may have good chemical stability, so that broad changes in eluent composition have no detrimental effects on the stationary phase. In addition, the pressure drop is usually lower than in beds packed with conventional particles, so that higher eluent flow rates can be used without incurring unduly high pressure drops [7].

We have examined the use of fibers which have been pre-assembled into a readily usable form i.e. yarns which in turn make up a fabric. Only woven fabrics are used in this work, but non-woven fabrics might also be useful sorbents. The fabric, when packed tightly in a column, retains the advantage of low pressure drops (typical of a fiber-based stationary phase). The fibers in the yarns are intertwined in both the warp (axial) and filling (radial) directions, which helps to make the stationary phase resistant to compaction at high pressure and flow rate. Derivatization of the fabric is readily accomplished, and can be easily carried out on a large scale using existing, cost-effective manufacturing techniques commonly associated with dyeing, crosslinking, or finishing of textiles yarns and fabrics.

The packing of fibrous stationary phases has previously been reported in the literature for a variety of configurations. These include: yarn or fiber wound in a spiral manner around a rod which is then packed into a column [8]; yarn or fiber in a parallel alignment which is then packed into a column longitudinally [6, 9]; randomly oriented fiber, yarn, fabric chunk and powder [10]; and ordered packing of disks of batting and fabric [7, 10, 11]. Our work differs from these previous studies in that the yarns in our fabrics have orientations that are both parallel and perpendicular to the flow direction.

Selective stepwise desorption, exploiting "on-off" chromatography [12], was found to be an efficient separation mode for the protein mixtures used in this work. In this mode, the feed mixture is first fed into the column in an eluent chosen so that all feed components are strongly retained. Then the composition of the eluent is changed in a series of steps so that only one feed component desorbs in each step. Ideally, each component is essentially unretained in its desorption step and emerges as a sharp band close to the void volume. There should be very little mixing between the feed components in the effluent; consequently, the plate count of the column becomes a relatively unimportant parameter in this mode of separation. Although the columns used here have modest plate counts (30–50 for a 50 cm column), they gave excellent separations of protein mixtures.

3 Experimental Details

3.1 Properties of Proteins

Properties of columns packed with viscose rayon fabric, diethyl aminoethyl cellulose (DEAE cellulose) fabric and sulfated cellulose fabric (derivatized in our laboratory) were studied through the separation and purification of BSA, IgG, insulin and β-gal. Serum albumin acts as a carrier of a number of substances in the blood, such as fatty acids; immunoglobulins G are the main class of antibodies raised by the immune system against bacteria, viruses, and toxins in the blood serum; insulin is a hormone protein which promotes the entry of glucose and amino acids from the blood into muscle and fat cells; and β-gal is the enzyme which hydrolyzes lactose to galactose and glucose [13–16].

We show that DEAE cellulose and sulfated cellulose fabric RSP are good anion- and cation-exchange chromatography materials, and that viscose rayon fabric RSP can be used in adsorption chromatography.

3.2 Chemicals

DEAE Cellulose: 2-(diethylamino) ethyl chloride is from Sigma Chemical Company (St. Louis, MO) and sodium hydroxide (AR) is from Mallinckrodt, Inc. (Paris, KY).
Sulfated Cellulose: The sulfuric acid (AR) used to prepare the sulfated cellulose was from Mallinckrodt.
Buffer Solution: Trizma buffer (Sigma), was used in the solvent to keep the pH at 7.2. Other buffers were 80 mM sodium acetate, NaAc, (AR, Mallinckrodt), and

80 mM acetic acid, HAc, (reagent, Fisher), which were mixed to give a buffer with pH of 4.7. 2 M Ammonium chloride, NH_4Cl, (AR, Mallinckrodt) with a pH of 4.5 is also used. A concentration of 5 mM disodium phosphate, Na_2HPO_4, (reagent, J. T. Baker Chemical, Phillipsburg, NJ) was used with 5 mM monopotassium phosphate, KH_2PO_4, (reagent, Matheson, Coleman and Bell Manufacturing Chemists, Norwood, OH) to give a buffer with a pH of 6.9; and 10 mM sodium perborate, $Na_2B_4O_7$, (Matheson, Coleman & Bell) was used to adjust the pH to 9.2.

Proteins: BSA (fraction V powder), anti-human IgG (whole molecule, developed in rabbit), insulin (from bovine pancreas) and β-gal (grade VIII, from *E. coli*) are all from Sigma. The concentration used in the LC studies was $10 \, g \, l^{-1}$ (solids concentration), except for insulin, which was used as a saturated solution. The proteins used were all dissolved in the eluent.

3.3 Stationary Phases

Viscose rayon, DEAE cotton and sulfated cotton were used as RSP. Cotton (TF #400 bleached cotton print test fabric) and viscose rayon (TF #215 filament viscose satin) are standard test fabrics from Testfabrics, Inc., Middlesex, NJ.

DEAE cotton was derivatized from the cotton fabric. The fabric was soaked in a 0.5 M solution of 2-(diethylamino) ethyl chloride at ambient temperature (25 °C) for 15 min, and squeezed through padded rollers (Crompton and Knowles Corp., Mauldin, SC) (two-dip-two-nip) to yield a wet pick-up of 100%. Then the fabric was immersed in 6 M NaOH solution for 45 min, followed by washing in flowing tap water until neutral pH was reached. The chemical reactions during the treatment are given in Fig. 1 [17]. The hydroxyl groups at

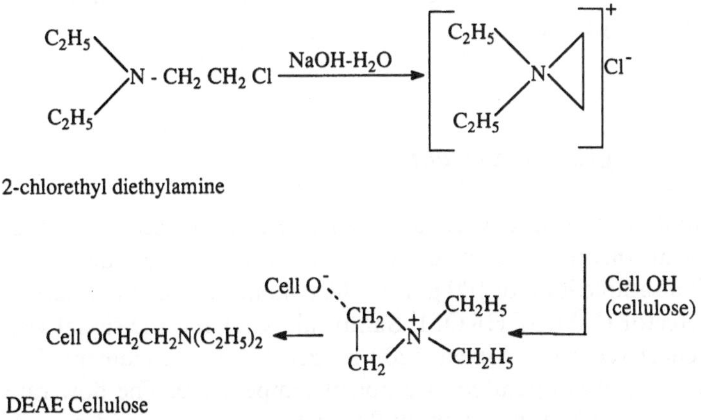

2-chlorethyl diethylamine

DEAE Cellulose

Fig. 1. Derivatization of DEAE-cellulose (after Rowland et al. [17])

the C-2 positions of the glucose residues have substantially more accessibility to the amine than do those at C-3 and C-6.

Sulfated cotton was also made from TF#400 bleached cotton print test fabric using an *in situ* technique. The fabric was first rolled and packed into the column. The column was capped with standard chromatography end fittings and washed with water. After washing, the eluent was changed from water to 6.0 M H_2SO_4 at a flow rate of 5 ml min^{-1} for 20 min. The column was then allowed to sit for 30 min, after which the eluent was changed to 1.0 M NaOH. Once the pH of the effluent reached 14, washing was continued for 5 min. The column was then rinsed with deionized water until the pH fell below 7. Sulfation was thus done at room temperature. The conditions used here were modified from the method given by Wadsworth and Daponte [18]. The chemical reactions can be simply expressed as below:

$$Cell - OH + H_2SO_4 \rightarrow Cell - OSO_3H \xrightarrow{NaOH} Cell - OSO_3Na$$

3.4 Column packing

The fabric was rolled tightly and then pulled into the column (10 mm i.d. × 500 mm, Supelco Inc., Belafonte, PA). Some columns were wet-packed (i.e. wetted with water during rolling), and others were dry-packed (i.e. rolled in dry form). Other fabric rolls were treated with 6 M NaOH (to promote shrinking) before packing.

These methods were used in order to select the packing technique for each material that gave the densest packing. In all cases, the overall porosity (as measured chromatographically) was in the range of 0.25–0.30, which is low compared to the 0.7–0.8 range characteristic of HPLC packings, indicating that these fabrics were tightly packed. The total surface area for underivatized cotton has been found [19] to be 30–45 $m^2 g^{-1}$ for pores larger than 60 Å, and 1–2 $m^2 g^{-1}$ for pores larger than 600 Å.

3.5 Liquid Chromatography System

Two pumps (Minipump, Milton Roy, Riviera Beach, FL) were used in parallel to pump the solvent through an injection valve (Rheodyne Model 7125, Berkeley, CA) with a sample loop of 100 µl into the column. The effluent passed through a VUV detector (VARI-CHROM, Sunnyvale, CA) and the signal was recorded using a chart recorder (Linear 1200 model, Linear Instrument Co., Irvine, CA). The column was operated at ambient temperature. The flow rate was 17.5 ml min^{-1}, which was the maximum flow rate of the combined pumps. The pressure drop of the overall LC system was about 400 psi at the highest flow

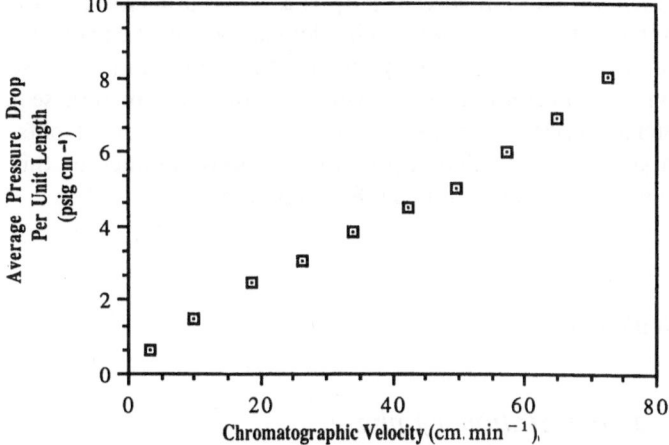

Fig. 2. Pressure drop as a function of chromatographic velocity for DEAE cellulose column. Temperature = 23 °C; column dimensions: 10 mm i.d. × 500 mm. The total porosity was taken to be 0.3

rates. A representative plot of pressure drop as a function of chromatographic velocity for DEAE cotton is shown in Fig. 2.

3.6 Stepwise Separation

The basis for the stepwise method is formalized as follows. Consider a binary feed mixture such that, in solvent A, component a was adsorbed but component b was unretained whereas in solvent B, component a was unretained but b was adsorbed. If the mixture of a and b were injected into the column using A as eluent, component b would elute while component a would be adsorbed. Switching the solvent from A to B would then elute component a, resulting in separation. The retention properties of proteins in several solvents is described by using " + " to signify retention, and " − " to signify effectively zero retention. In general, for n proteins and n solvents, an $n \times n$ matrix, $(a_{ij})_{n \times n}$, could be obtained as shown in Fig. 3, where the row variable, i, represents different proteins and the column variable, j, represents different

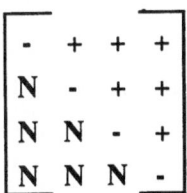

Fig. 3. Requirements for selective stepwise desorption schedule. The entry " − " implies that the relevant component is effectively unretained under the specified conditions; " + " implies that the component is strongly bound under these conditions. "N" is used when the retention is irrelevant to the stepwise desorption schedule

solvents. If protein i does not elute in solvent j, a_{ij} is a " + "; otherwise, it is a " − ". If the matrix can be written as one with all its elements above the principal diagonal being " + " and the elements at the principal diagonal being " − " by changing the orders of rows and columns only, as in Fig. 3, the mixture of these n proteins can be completely separated in a stepwise fashion by these n solvents. The separation of a mixture of four different proteins has been obtained by this method on both DEAE and sulfated cellulosic RSP columns.

4 Results and Discussion

4.1 DEAE Cotton as the Stationary phase

Adsorption properties of BSA, IgG, insulin and β-gal were studied using the DEAE cellulose RSP with water (pH 5.5), NaCl and different concentrations of Trizma buffer at pH 7.2 as eluents. The results are summarized in Table 1.

Using deionized water (pH 5.5) as an eluent, all the test proteins were held up on the column. All the retained proteins were rapidly desorbed in NaCl solutions and elute as a sharp peak. The eluent flow rate is 17.5 ml min^{-1} throughout. Figure 4 shows BSA adsorption in water and desorption in 2 M NaCl, due to the anionic exchange properties of the DEAE cellulose. The mechanism can be expressed as follows (Fig. 5).

If the solvent is changed from water to 25 mM Trizma buffer, and the pH of the solution is increased from 5.5 to 7.2, many proteins will be more negatively charged than in pure water. This promotes their adsorption onto an anion-exchange stationary phase. In addition, using the Trizma buffer also increases the ionic strength of the eluent. Buffer anions will then compete more effectively with the proteins' anions for the positively charged fixed sites of the bed. This competitive effect will also moderate protein adsorption. Since proteins have different isoelectric points and different structures, they are affected to different

Table 1. Influence of eluent properties on adsorption of sample proteins on DEAE cotton columns. " + " implies strong retention, " − " implies lack of (or very weak) retention

Eluent/Protein	IgG	BSA	Insulin	β-Galactosidase
DI Water (pH 5.5)	+	+	+	+
25 mM Trizma buffer (pH 7.2)	−	+	+	+
50 mM Trizma buffer (pH 7.2)	−	−	−	+
2 M NaCl	−	−	−	−

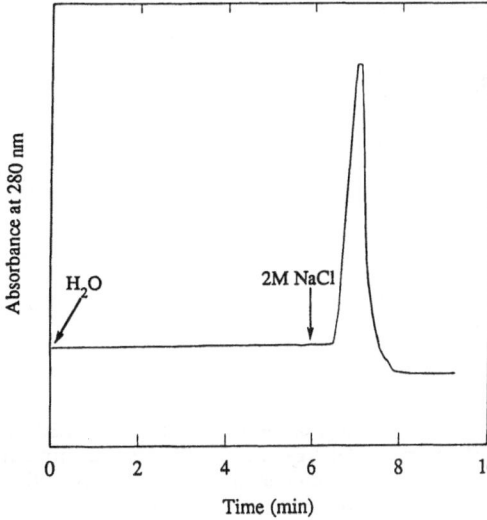

Fig. 4. Stepwise adsorption and desorption of BSA to DEAE-cellulose. BSA is adsorbed on to the bed from pure water, and desorbed in 2 M NaCl. Ambient temperature; flow rate = 17.5 ml min^{-1}; chart speed = 1 cm min^{-1}. Column dimensions: 10 mm i.d. × 500 mm. Sample volume = 100 µl; protein concentration = 10 g l^{-1}

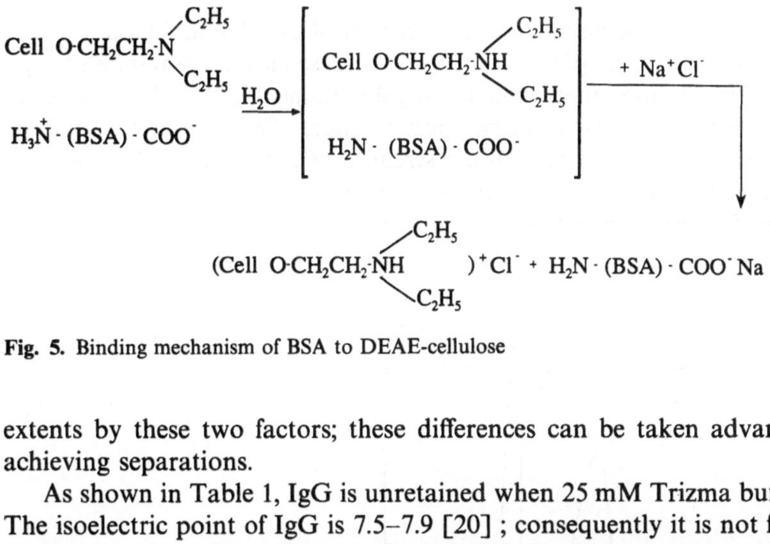

Fig. 5. Binding mechanism of BSA to DEAE-cellulose

extents by these two factors; these differences can be taken advantage of in achieving separations.

As shown in Table 1, IgG is unretained when 25 mM Trizma buffer is used. The isoelectric point of IgG is 7.5–7.9 [20] ; consequently it is not far from its isoelectric point in the Trizma buffer (pH 7.2). Its electrostatic interactions with the stationary phase are therefore weak and even a small amount of other anions can prevent it from binding. By contrast, the addition of 25 mM buffer has a totally different effect on BSA, which has an isoelectric point of 4.9 [20]; the electrostatic attraction between the protein and the adsorbent becomes the dominant factor. Furthermore, the BSA will elute from the column when 2 M NaCl (in deionized water, pH 5.5) solution is used. Figure 6 shows the separation of IgG and BSA mixture with the buffer-NaCl system. The protein mixture is first injected into the column in the buffer. After the IgG emerges, 2 M NaCl is used to elute the BSA. As shown, separation is achieved in 3 min.

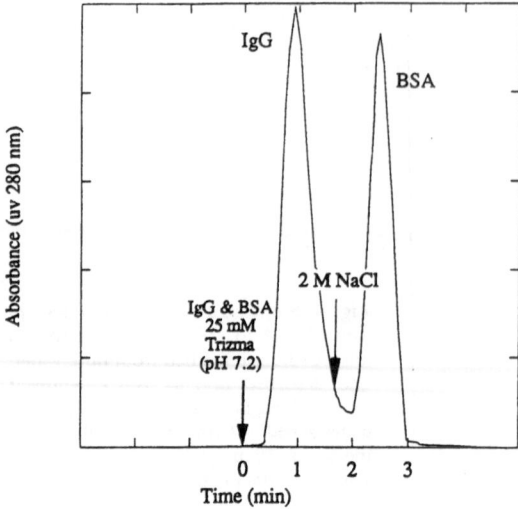

Fig. 6. Separation of IgG and BSA on DEAE-cellulose. Stepwise desorption conditions given on figure. All other conditions as in Fig. 4

To examine the importance of ionic strength of the eluent, 50 mM Trizma buffer was used at the same pH of 7.2. At this increased ionic strength, BSA, IgG and insulin were all unretained while the β-gal remained adsorbed. The electrostatic interactions of the first three components were effectively nullified at this higher ionic strength, while the binding strength of β-gal was initially so high that it still remained charged. Using this fact, a good separation of insulin and β-gal is obtained within 5 min as shown in Fig. 7.

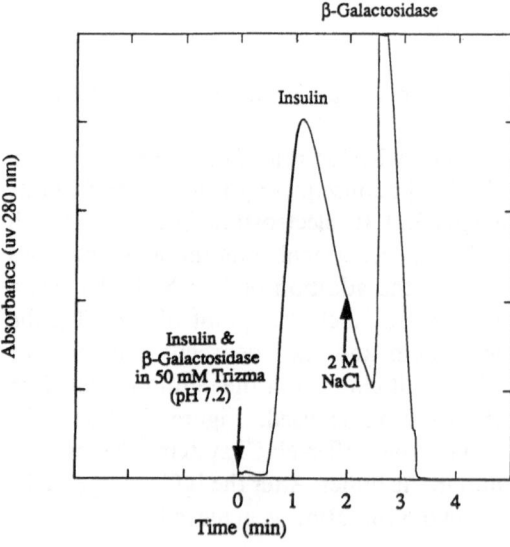

Fig. 7. Separation of insulin (saturation concentration) and β-gal on DEAE-cellulose. Stepwise desorption conditions given on figure. All other conditions as in Fig. 4

4.2 Sulfated Cotton as the Stationary Phase

Different eluent pHs and salts were used to test the cation exchange capabilities of the column with respect to the test proteins. The results are summarized in Table 2. It was also found that insulin and β-gal were retained on the sulfated cotton even in 2 M NaCl or 2 M NH$_4$Cl solution, probably reflecting the high acidity of the sulfate group on the cellulose structure of the cotton, with the cellulose having a high degree of substitution.

BSA, IgG, and β-gal were eluted from the sulfated cotton bed in 5 mM Na$_2$HPO$_4$–KH$_2$PO$_4$ buffer. The cation effect of the buffer was low since the eluent was dilute compared with 2 M NaCl and 2 M NH$_4$Cl, but the pH of the solution was increased from 5.5 to 6.9, indicating that the influence of the ionic strength predominated over the electrostatic interactions. When the pH was further increased to 9.2 (with 10 mM Na$_2$B$_4$O$_7$), all four proteins became effectively unretained.

The sensitivity of proteins to pH changes in ion-exchange chromatography can be beneficial since the adsorbed proteins can be washed out without requiring strong salt solution. Purification of proteins from salt buffers is therefore avoided. Instead of using simple electrolyte ions to substitute for the proteins on the ion-exchange column, the pH is adjusted to neutralize the proteins' surface charges. Because the amine group is a weakly basic group, it is more sensitive to pH changes than is the carboxylic acid group. Thus pH adjustment has a greater effect on protein cation-exchange than on anion exchange. Consequently, a cation-exchange column intended to separate and purify proteins can also be used to desalt them, since pH adjustment is readily achieved by only a slight change in ionic strength of the eluent. Volatile buffering agents that can be easily removed, such as ammonium hydroxide or acetic acid, can be used.

The changes in operating conditions motivated above, and described in Table 2, achieved the separation of the 4-component test mixture in 9 min, as

Table 2. Influence of eluent properties on adsorption of sample proteins on sulfated cotton columns. Symbols as in Table 1

Eluent/Protein	BSA	IgG	β-Galactosidase	Insulin
DI Water (pH 5.5)	+	+	+	+
80 mM NaAc-HAc (pH 4.7)	−	+	+	+
2.0 mM NH$_4$Cl (pH 4.5)	−	−	+	+
5 mM Na$_2$HPO$_4$ KH$_2$PO$_4$ (pH 5.9)	−	−	−	+
10 mM Na$_2$B$_4$O$_7$ (pH 9.2)	−	−	−	−

shown in Fig. 8. BSA was first removed by bringing the eluent pH (4.7 for the NaAc-HAc buffer) close to its isoelectric point. Then the high ionic strength of the NH_4Cl solution caused the IgG to desorb. These first two desorption steps were driven by changes in ionic strength, analogous to the separation of IgG on DEAE-cellulose described above. After this, the other two proteins-β-gal and insulin-were separated by pH effects, since the ionic strength in both of these steps was low.

Comparing Table 1 with Table 2, all four proteins studied can be adsorbed on either DEAE or sulfated cottons in deionized water (pH 5.5). It is not necessary to have net positive nor net negative charges of proteins for them be adsorbed on cation- or anion-exchange RSP.

When Trizma buffer was used, all the proteins adsorbed on the sulfated cotton bed were readily desorbed and the column no longer had cation-exchange properties unless subsequently washed with a strong acid solution. In solution, Trizma buffer has a large tri(hydroxymethyl) aminomethane ion, $H_3N^+C(CH_2OH)_3$. The CH_2OH group is electronegative, and the electron-withdrawing inductive effect of the three CH_2OH groups gives the ion a strong attraction to the sulfate groups of the bed. Further, the bulky character of the four big groups around the nitrogen atom shields the charge from other positive ions. We believe that inductive and steric effects are both involved in causing the Trizma ion to adsorb strongly on the cation-exchange bed, and thus preventing the (further) adsorption of proteins. Four parameters influence protein adsorption on an ion exchange bed: basicity (K_b) or acidity (K_a) of the adsorbent; isoelectric point and structure of the protein; pH of the eluent; and its ionic strength. The first two factors cannot be changed once the adsorbent has been

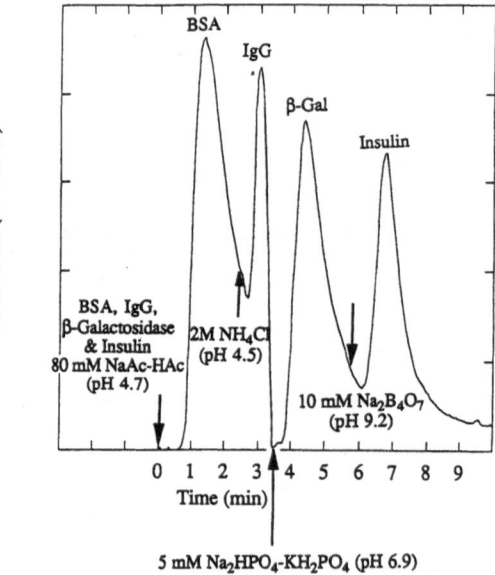

Fig. 8. Separation of a four-protein mixture on a sulfated-cotton column. Stepwise desorption conditions given on figure. All other conditions as in Fig. 4

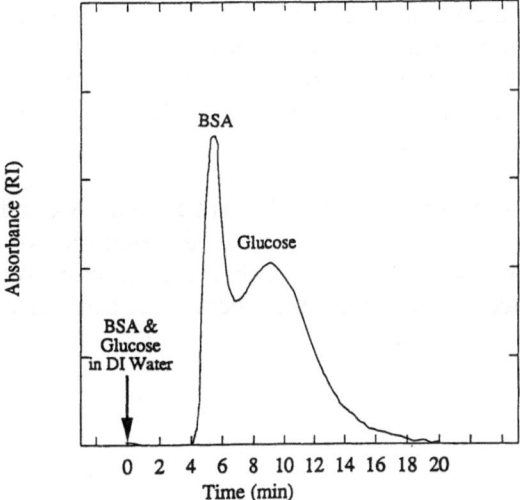

Fig. 9. Separation of BSA and glucose in water on viscose rayon. Flow rate = 1.0 ml min^{-1}; chart speed = 6 cm h^{-1}. All other conditions as in Fig. 4

chosen. However, the latter two factors are adjustable, allowing many possible strategies for protein separation.

4.3 Viscose Rayon as the Stationary Phase

As has been discussed, effective ion-exchange separations can be obtained by suitably manipulating the properties of the fabric's functional groups. Since there are so many commercially available textile fabrics having different chemical and physical structures, fabric-based chromatography columns with a variety of properties can be envisaged. An example is viscose rayon, which is a cellulosic fiber derived from wood. Because of the cellulosic structure of the RSP, glucose is more strongly retained on viscose rayon columns than are proteins. Figure 9 depicts the separation of BSA from glucose. Since BSA is a big molecule (MW = 67 000), it is essentially excluded from the viscose rayon and comes out earlier. Glucose, being much smaller (MW = 180), penetrates the intraparticulate pore space, and is retained for a longer period.

5 Conclusions and Future Work

The use of polymeric continuous stationary phases for preparative chromatography has been investigated. Preliminary experimentation indicates that these systems have several attractive features. The relatively inexpensive starting material (fabric) is encouraging; further, RSP can be mass-produced to within close tolerances using textile machinery. These stationary phases are chemically and mechanically stable at high flow rates. Further work includes measuring loadings and recoveries, so that quantitative comparisons of such columns may be made on the basis of performance variables such as throughput and yield with conventional chromatographic columns.

Acknowledgments. This work was supported by NSF Grants EET 8907304 and BCS 8912150. We would like to thank Karen Kohlmann, Richard Hendrickson, George Tsao, and Robert Dean for their detailed scrutiny of the manuscript. We would also like to acknowledge Richard Hendrickson's experimental assistance.

6 References

1. Reisman HB (1988) Economic analysis of fermentation processes. CRC Press, Boca Raton
2. Sofer GK, Nystrom L-E (1989) Process chromatography. Academic Press, San Diego
3. Knight P (1989) Bio/Technology 7:777
4. Eble JE, Grob RL, Antle PE, Snyder LR (1987) J Chromatogr 384:254
5. Gibbs SJ, Lightfoot EN (1986) Ind Eng Chem Fundam 25:490
6. Ding H, Yang M-C, Schisla D, Cussler EL (1989) AIChE J 35:814
7. Osawa AE, Cooney CL (1989) Paper No. 3, MBTD Division, 198th American Chemical Society National Meeting, Sept 10–15, 1989, Miami Beach, FL
8. Sharma SC, Fort JT (1973) J Colloid Interface Sci 43:36
9. Kiso Y, Jinno K, Nagoshi T (1986) J High Res Chromatogr Commun 9:763
10. Rowland SP, Wade CP, Bertoniere NR (1984) J Appl Polym Sci 29:3349
11. Bertoniere NR, King WD (1989) Textile Res J 59:114
12. Tiselius A (1955) Angew Chem 67:245
13. Kwapinski JBG (1972) Methodology of immunochemical and immunological research. Wiley-Interscience, New York
14. McGarry JD (1986) In: Devlin TM (ed) Textbook of biochemistry with clinical correlations. John Wiley, New York
15. Stryer L (1981) Biochemistry. Freeman, New York
16. Watson JD, Tooze J, Kurtz DT (1983) Recombinant DNA. Freeman, New York
17. Rowland SP, Roberts EJ, Wade CP (1969) Textile Res J 39:530
18. Wadsworth LC, Daponte D (1985) In: Nevell TP, Zeronian H (ed) Cellulose chemistry and its applications, Ellis Horwood, Chichester, West Sussex, UK
19. Ladisch CM, Yang Y, Velayudhan A, Ladisch MR (1992) Textile Res J 62:36
20. Righetti PG, Caravaggio T (1976) J Chromatogr 127:1

Author Index Volumes 1–49

Ohshima, T., Soda, K.: Biochemistry and Biotechnology of Amino Acid Dehydrogenases. Vol. 42, p. 187
Okabe, M. see Aiba S. Vol. 7, p. 111
Olson, N. F. see Finocchiaro, T. Vol. 15, p. 71
Onken, U., Liefke, E.: Effect of total and partial Pressure (Oxygen and Carbon Dioxide) on Aerobic Microbial processes. Vol. 40, p. 137

Pace, G. W., Righelato, C. R.: Production of Extracellular Microbial. Vol. 15, p. 41
Parisi, F.: Energy Balances for Ethanol as a Fuel. Vol. 28, p. 41
Parisi, F.: Advances in Lignocellulosic Hydrolysis and in the Utillization of the Hydrolyzates. Vol. 38, p. 53
Parulekar, S. J., Lim, H. C.: Modelling, Optimization and Control of Semi-Batch Bio-reactors. Vol. 32, p. 207
Pécs, M. see Nyeste, L. Vol. 26, p. 175
Phillipson, J. D. see Anderson, L. A. Vol. 31, p. 1
Pitcher Jr., W. H.: Design and Operation of Immobilized Enzyme Reactors. Vol. 10, p. 1
Potgieter, H. J.: Biomass Conversion Program in South Africa. Vol. 20, p. 181
Poweigha, T. see Kleinstreuer, C. Vol. 30, p. 91
Primrose, S. B.: Controlling Bacteriophage Infections in Industrial Bioprocesses. Vol. 43, p. 1
Prokop, A., Rosenberg, M. Z.: Bioreactor for Mammalian Cell Culture. Vol. 39, p. 29

Quicker, G. see Schumpe, A. Vol. 24, p. 1

Radlett, P. J.: The Use Baby Hamster Kidney (BHK) Suspension Cells for the Production of Foot and Mouth Disease Vaccines. Vol. 34, p. 129
Radwan, S. S., Mangold, H. K.: Biochemistry of Lipids in Plant Cell Cultures. Vol. 16, p. 109
Ramasubramanyan, K., Venkatasubramanian, K.: Large-Scale Animal Cell Cultures: Design and Operational Considerations. Vol. 42, p. 13
Ramakrishna, D.: Statistical Models of Cell Populations. Vol. 11, p. 1
Rapoport, A. J. see Beker, M. J. Vol. 35, p. 127
Reardon, K. F. see Friehs, K. Vol. 48, p. 53
Reese, E. T. see Faith, W. T. Vol. 1, p. 77
Reese, E. T., Mandels, M., Weiss, A. H.: Cellulose as a Novel Energy Source. Vol. 2, p. 181
Řeháček, Z.: Ergot Alkaloids and Their Biosynthesis. Vol. 14, p. 33
Rehm, H.-J., Reiff, I.: Mechanism and Occurrence of Microbial Oxidation of Long-Chain Alkanes, Vol. 19, p. 175
Reiff, I. see Rehm, H.-J. Vol. 19, p. 175
Reinhard, E., Alfermann, A. W.: Biotransformation by Plant Cell Cultures. Vol. 16, p. 49
Reiser, J., Glumoff, V., Kälin, M., Ochsner, U.: Transfer and Expression of Heterologous Genes in Yeasts Other Than *Saccharomyces cerevisiae.* Vol. 43, p. 75
Reuveny, S. see Shahar, A. Vol. 34, p. 33
Richardson, T. see Finocchiaro, T. Vol. 15, p. 71
Righelato, R. C. see Pace, G. W. Vol. 15, p. 41
Rivera, S. L. see Karim, M. N. Vol. 46, p. 1
Roberts, M. F. see Anderson, L. A. Vol. 31, p. 1
Roels, J. A. see Harder, A. Vol. 21, p. 55
Rogers, P. L.:. Computation in Biochemical Engineering. Vol. 4, p. 125
Rogers, P. L., Lee, K. L., Skotnicki, M. L., Tribe, D. E.: Ethanol Production by Zymomonas Mobilis. Vol. 23, p. 37
Rolz, C., Humphrey, A.: Microbial Biomass from Renewables: Review of Alternatives. Vol. 21, p. 1
Rosazza, J. P. see Smith, R. V. Vol. 5, p. 69
Rosenberg, M. Z. see Prokop, A. Vol. 39, p. 29
Rosevear, A., Lambe, C. A.: Immobilized Plant Cells. Vol. 31, p. 37
Ryhiner, G. B. see Heinzle, E. Vol. 48, p. 79

Sahm, H.: Anaerobic Wastewater Treatment. Vol. 29, p. 83
Sahm, H.: Metabolism of Methanol by Yeasts. Vol. 6, 77
Sahmn, H.: Biomass Conversion Program of West Germany. Vol. 20, p. 173
Sambanis, A. see Grampp, G. E. Vol. 46, p. 35

Subject Index